Mining and Indigenous Lifeworlds in Australia and Papua New Guinea

Mining and Indigenous Lifeworlds in Australia and Papua New Guinea

Edited by Alan Rumsey and James Weiner

Sean Kingston Publishing
www.seankingston.co.uk
Wantage

This edition published in 2004 by

Sean Kingston Publishing

www.seankingston.co.uk

57 Orchard Way, Wantage, Oxon, OX12 8ED, UK

British Library Cataloguing-in-Publication Data

A catalogue record for this book is available from the British Library

Printed by Lightning Source

Paperback ISBN 0–9545572–3–9

Hardback ISBN 0–9545572–4–7

CONTENTS

ACKNOWLEDGEMENTS

This volume originated in the conference 'From Myth to Minerals: Place, Narrative, Land and Transformation in Australia and Papua New Guinea', held at the Australian National University (ANU) in July 1997. The conference was funded primarily by a Wenner-Gren International Foundation for Anthropological Research conference grant. We would like to thank Sydel Silverman and Laurie Obbink of Wenner-Gren for their help and support. Additional financial help was provided by the Resource Management in Asia Pacific Project of the Research School of Pacific and Asian Studies, ANU.

Over four days, the conference heard more than thirty papers and was attended by approximately 150 people. It was the largest single conference ever designed specifically to bring Aboriginalists and Melanesianists together to discuss common contemporary theoretical and ethnographic issues. In addition to *Mining and Indigenous Lifeworlds*, another volume of papers from the conference, *Emplaced Myth*, has also appeared with the University of Hawai'i Press.

For their contributions and support, all of these people, including those original invitees whose papers for one reason or another we could not include in either of these volumes, deserve our deepest thanks and collegial gratitude.

Dianna Kovacs, Fay Castles and Ria van de Zandt of the Research School of Pacific and Asian Studies provided invaluable logistic and editorial support during and after the conference. Mr John Hudson of Burton and Garran Hall helped ensure that our conference visitors were comfortably housed and provided the venue and catering for the conference.

Margaret Forster provided editorial services during the preparation of this manuscript. She was supported by an Australian Research Council Large Grant.

AUTHOR AFFILIATIONS AND INSTITUTIONS

Ernst, T.M.: School of Humanities and Social Sciences, Charles Sturt University, Wagga Wagga, New South Wales

Gardner, Don: Department of Archaeology and Anthropology, The Faculties, Australian National University (ANU), Canberra

Jorgensen, Dan: Department of Anthropology, University of Western Ontario

Keen, Ian: Department of Archaeology and Anthropology, The Faculties, ANU, Canberra

Kirsch, Stuart: Department of Anthropology, University of Michigan

McIntosh, Ian: Cultural Survival, Cambridge, Massachusetts

Merlan, Francesca: Department of Archaeology and Anthropology, The Faculties, ANU, Canberra

Robinson, Michael: Department of Anthropology, University of Western Australia, Perth

Rumsey, Alan: Department of Anthropology, Research School of Pacific and Asian Studies, ANU, Canberra

Sagir, Bill F.: Department of Anthropology, Research School of Pacific and Asian Studies, ANU, Canberra, and Department of Anthropology and Sociology, University of Papua New Guinea, Port Moresby

Strang, Veronica: Department of Anthropology, University of Wales, Lampeter

Trigger, David: Department of Anthropology, University of Western Australia, Perth

Wardlow, Holly: Department of Anthropology, University of Iowa

Weiner, James F.: Department of Anthropology, Research School of Pacific and Asian Studies, ANU, Canberra

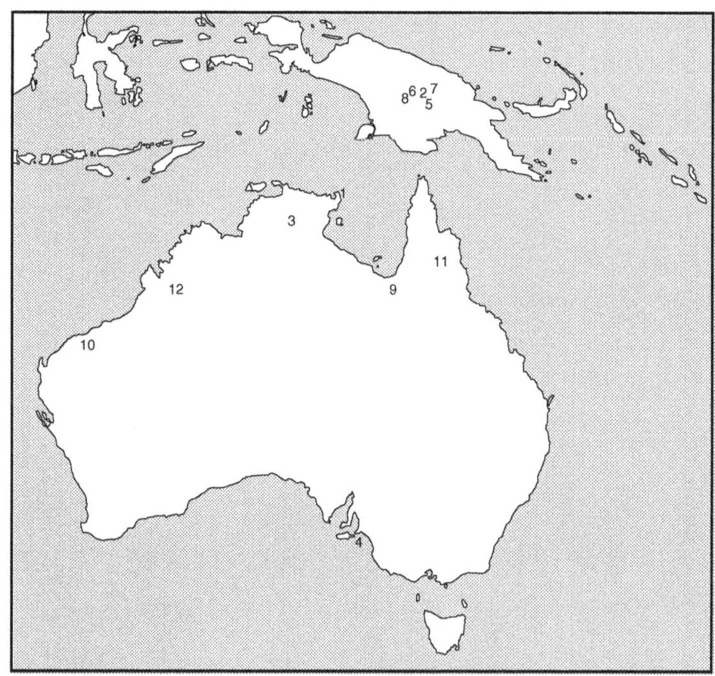

1. Yirkalla (Yolngu)
2. Mt. Kare (Huli)
3. Coronation Hill (Jawon)
4. Hindmarsh Island (Ngarrindjeri) [Bridge development]
5. Lake Kutubu (Fasu, Foi)
6. Nena (Telefolmin and Miyanten)
7. Porgera (Ipili)
8. OK Tedi
9. Century Zinc (Waanyi, Garrwa)
10. Woodside Petroleum (Pilbara Coast)
11. Red Dome (Mitchell River headwaters)
12. Noonkanbah (Fitzroy Valley)

Map 1: Sites of mining and development that are discussed in this volume.

1
INTRODUCTION: DEPOSITINGS

James Weiner

> There is no mother tongue, only a power takeover by a dominant language within a political multiplicity. Language stabilizes around a parish, a bishopric, a capital. It forms a bulb. It evolves by subterranean stems and flows, along river valleys or train tracks; it spreads like a patch of oil. [Deleuze and Guattari 1983:7]

A great deal has been written on the political, economic, environmental and developmental aspects of mineral resource extraction in areas inhabited and/or controlled by indigenous people. In the Australia-South Pacific region, several volumes have been devoted to issues surrounding the confrontation between indigenous people, states and resource extraction companies. Important collections include those edited by Berndt (1982), Connell and Howitt (1991), Henningham and May (1992), Howitt et al. (1996), Toft (1997), Larmour (1997), Brown and Ploeg (1997), and Banks and Ballard (1997) (the latter devoted specifically to analysing the implications of the Ok Tedi settlement). These studies are valuable in that they give a very broad and multifaceted view of a complex global process, namely large-scale mining in the indigenous world. However, from an anthropological perspective, these studies leave largely unexamined and unanalysed the nature of knowledge systems and the culturally distinctive epistemological and discursive processes *within* indigenous societies in these contexts. With but few exceptions, such as Clark (1994), Biersack (1995a) and Guddemi (1997), the absence of sophisticated understandings of indigenous motivations and cultural frameworks only seems to loom larger and more critical in the current period of extensive mining in indigenously populated areas.

This gap in previous work was a major stimulus for the 1997 conference, 'From Myth to Minerals: Person, Narrative, Land and

Transformation in Australia and Papua New Guinea' sponsored by the Wenner-Gren Foundation for Anthropological Research and held at the Australian National University. This volume of papers by a number of the anthropologists who attended that conference, and several who were invited to contribute to it afterwards, attends specifically to the issue of how the presence of mineral resources and the mining companies become events of mythological and cosmological significance in the lifeworlds of indigenous people.

The gender of the gilt

We begin by appealing to the process of 'depositings' to draw attention to the material consequences of lived being – the constant shedding of bodily exuviae and organs which consequently attain their own reproductive and destructive powers (and it is important to understand at the outset how intimately life and death are enfolded by the same sexual and reproductive processes). The relationship between human being and action and land is one in which a dynamic and reciprocal transfer of substance and material is integral to it. The intimacy between human beings and their emplacement on land begins with the imprint this ongoing substantial transfer leaves on the bodies of both. One of our main goals in this volume is to describe a process by which a local, embodied regime of corporealized territoriality takes over and is taken over by the kinds of radical, large-scale alteration of the land and landscape that accompany major mineral extraction projects in Australia and Papua New Guinea.

But there is a second dimension to this corporeal relation between the human body and the body of the earth. In Papua New Guinea and Australia, the transformation of human and ancestral action and substance into features of the landscape is mediated by highly specific and formalised genres of language such as myth and song. In the same way, however, that we insist on a dynamic and constitutive relation between human and territorial corporeality, we also eschew the appeal to myths as congealed narrative and focus more on *mythopoiesis*, the causing to come about through special mythic-narrative techniques. These techniques are part of the discursive and verbal repertoire of agents more generally in these settings. An appeal to mythopoiesis as a knowledge-formation process draws attention to the active, rather than merely expressive, ways in which agency is established and made visible through such techniques.

The ethnographic accounts of verbal expression in Papua New Guinea and Australia indicate that speech and the act of inscription – whether of the human body, the ground itself, or any artefact or representational surface – are inseparable. Furthermore, the actions of ancestral beings were creative both of the physical particularities of the landscape and its onomastic properties. We draw attention to the particular quality of Papua New Guinean and Australian landscapes: that they are parcelled into discrete sites through these acts of naming.

The reproductive and procreative dimensions of these acts of inscription and naming are also salient in both regions. In the case of the great Wawilak and Djangkawau cycles of Northern Territory, the genitals of the creator beings either pierced or marked the land themselves, or these organs were severed and the remnants became landmark features. A similar process of sexual marking was found among the Marind-anim of the south coast of what is now Irian Jaya (see Weiner 1995a).

The discovery of valuable mineral deposits in Papua New Guinea and Australia is often interpreted by indigenous people as further evidence of such processes of sexualised marking. At both Mount Kare (Clarke 1994) and Lake Kutubu (Weiner 1995a) in Papua New Guinea, the presence of minerals is linked with sexual processes of human reproduction. Furthermore, with respect to the linking of gold deposits with the python in the Porgera mine in Enga Province, Biersack maintains that these equations index the regenerative capability of the body (Biersack, pers. comm.). We are confronted with a form of inscription of the landscape, which, as is the case in Australia, is also a process of embodiment: the land is a body and its regenerative capabilities are the essence of fertility (cf. Rose 2001). Local people are thus likely to interpret the negative consequences of mineral extraction – pollution, the disappearance of vegetative cover, human sickness – as dysfunctions of an organism rather than of an inert environment (see also Strang, this vol.). For example, the red colour of the Porgera River caused by mine tailings is seen as a form of menstrual pollution, greatly feared by men of this region of Papua New Guinea (Biersack, pers. comm.).

A similar process is found in Australia, most notably in the case of the creator being Bulardemo at Coronation Hill in Northern Territory, as Merlan describes in her chapter for this volume. When gold deposits were found at Coronation Hill, the Jawoyn people interpreted these as residues of Bulardemo's ancestral body, and senior custodians of the site warned of dire consequences to the fabric of the cosmos should these deposits be disturbed, for example, through explosive blasting. Similarly, Jorgensen notes that in the vicinity of the Nena mine in West Sepik Province, Papua New Guinea:

> Local people now make a general association between geologically anomalous locations (marked by discoloured streams, different forms of rock, stunted flora, and so on) and the presence of valuable minerals, and many such locations were already viewed as sacred or powerful sites (*amemtem*), of which Nena Mountain is one. [Jorgensen, this vol:88]

Taussig has made a general observation in the face of these indigenous responses to the transformation of the landscape in his early powerful study of the indigenous history of mining in South America:

> ... the devil represents not merely the deep-seated changes in the material conditions of life but also the changing criteria in all their dialectical turmoil of truth and being with which those changes are associated – most especially the radically different concepts of creation, life, and growth through which the new material conditions and social relations are defined (Taussig 1980:17).

Taussig's point is in many respects consonant with that of Sahlins (1981, 1985): It is through the uses to which mythpoietic thought is put that we see most clearly the conjunctural contact between largely incommensurate cultural systems. This framework explicitly informs the chapters by Ernst and Merlan in this volume, and is implicit in all the others, in particular that of McIntosh. However, the contributors to *Mining and Indigenous Lifeworlds* find the conflictual nature of this mythopoietic conjuncture more salient than its totalising structural features. Both Merlan and Trigger and Robinson in their chapters discuss how the inherent creativity evident in the Aboriginal perception of ancestral activity clashes, often in dramatic court cases, with the markedly divergent epistemic framework of the secular west, in which this type of creativity is viewed as an exclusively human attribute. Hence, in this volume, we include in our view those continuing mythopoietic evenamental contexts that serve to erode the sharpness of this border, to which we now turn.

The emergence of landowner politics

A significant development in mineral resource extraction, particularly in Papua New Guinea, is the number of major projects that are located in areas that seem to be interstitial to different ethnic, linguistic and cultural groups. The mines at Nena (described by Gardner and Jorgensen in this volume) and Mount Kare (described by Wardlow in this volume), the oilfield at Gobe, the oil pipeline in the Southern Highlands and Gulf Provinces, all are foci of competition over ownership and jurisdiction by different named groups. But before we conclude that this is a curious accident of mineral geography, at least four observations must be made.

First, the Papua New Guinea government localises mining projects through the delineation of project development license areas. These areas are drawn with reference to the graticular units of fixed dimension on official maps. The area delineated thus invariably includes the immediate sites of the project, but is not restricted to them. Once delineated, all groups owning land within the project development area are considered project landowners. Thus, in the Kutubu Petroleum Development License area 2 (PDL-2) (described in Sagir's chapter in this volume), although only the Fasu actually own land upon which well-heads are located, some clans of the neighbouring Foi and Huli people whose land lies within the PDL are also included as landowners.

Second, partially as a result of these inclusive geopolitical demarcations, any site of major resource development is inevitably going to be a focus of competition by those groups who can claim some proprietary interest in it. In Papua New Guinea, as is the case throughout the world of indigenous subsistence cultivators and hunters and gatherers, different categories of kin have different rights to particular tracts of land and the resources in it, even if these kin are not the 'owners' in a strict sense. In both Australia and Papua New Guinea, the establishment of large-scale financial and business opportunities in remotely situated mining enclaves stimulates a return to one's home territory by those who had previously emigrated. In this volume, the two chapters by Jorgensen and Gardner both deal with the same topic – the dispute between the Telefolmin and Miyanten people over proprietorship of the mine at Nena in the West Sepik Province of Papua New Guinea. Both Jorgensen and Gardner worked as consultants on a team charged with determining the cultural and historical grounds for the different claims made by these groups, Jorgensen with the Telefolmin and Gardner with the Miyanten. We juxtapose these two chapters in this volume to show not only how extraordinarily complex such mytho-historical narratives of attachment and ownership are, but to underline the anthropologist's role in sharpening and deepening our understanding of how incommensurable these contrasting accounts can be made to appear.

Third, in both Papua New Guinea and Australia, it is not the boundary between delineated tracts that people articulate so much as a series of specific sites. These sites are linked in a variety of ways with other sites. People are more focussed on the tracks or paths between sites than they are on the borders between more encompassing territorial blocks. Any given site can be the node of multiple claims, obligations, rights and responsibilities. In his groundbreaking study, reprinted in this volume, Jorgensen observes that in the Ok Tedi region:

> The announcement that minerals had been discovered at Fubilan was greeted with enthusiasm, a reaction that owed some of its force to the peculiarities of the mine site's location. Situated almost due west of Telefolmin, and near the Wopkaimin hamlet of Bultem, Fubilan enjoys a special place in Telefol mythology as the site of the Land of the Dead (Bagelam) established by Afek. When Afek killed her younger brother, she sent him along an underground track to Fubilan, where he made a place for the dead and created the source of shells (*bonang*) and the stone *fubi* adzes (hence Fubilan) that were the principal items of traditional wealth. This, the mythology of Afek not only linked various peoples in a genealogy of her descendants: it also mapped relations among ancestors, the living, and wealth at Fubilan. Given the ritual primacy of Telefolip and the role of Telefolmin as custodians of Afek's legacy, the government's decision to recognise Wopkaimin as Fubilan's landowners generated consternation and resentment. [Jorgensen, this vol.:73]

Moreover, as Kirsch and Strang draw attention to in this volume, indigenous people living downstream of major mining projects, even at some distance, suffer the impact of environmental degradation caused by water-borne mineral effluvium. Not only does this make them "stakeholders" in the project's total effects, it demonstrates that no single mining project can be considered to have an exclusively "local" placement – they are all regional in their impact, involvement and consequences.

The chapters by Ernst, Gardner, Jorgensen, Merlan and Sagir focus on a fourth important phenomenon of conjunctural relations between companies, governments and indigenous people in mining situations: the attempt by company and government representatives to make an administrable, clearly defined 'entity' out of social groupings that are in anthropological terms contingent and fluid. The permeable boundaries of land-controlling units in both Papua New Guinea and Australia have, in the first place, been long associated with particular environmental conditions that put a premium on the flexibility of group membership. Strehlow (1970) and Peterson (1986) have been among those who have argued for this flexibility as a core feature of band organisation in the Northern Territory of Australia, while Modjeska has persuasively linked production and lineage form in one Central Highlands society of Papua New Guinea (1982). Gardner thus argues for Telefol and Miyanten people of the Nena mine area:

> ... the [local group] identity of individuals and groups is not idle or irrelevant to practice, but ... is always contingent upon discursive context, and is therefore liable to be contested; even when there is no contestation, it is not because identity has transcended such contexts, but because, *pro tem* and for the purposes to hand, a consensus has been achieved. [Gardner, this vol.:113-14]

John Burton's appeal to what he calls the 'cadastral' landscape is appropriate here. Drawing on the established legal sense of this term, Burton intends by it: 'the end result of dealings in land; it is the cumulative record of legal decisions – arising from customary or state law – in the formation of the cultural landscape' (1991:197). The point is that if we locate an important dimension of the power of place in the narrative resources of these societies, then we must consider how readily such narrative techniques appropriate and make contact with analogous Western techniques for doing the same. As Filer has convincingly argued in the Papua New Guinea context, the issue of land ownership is largely an artefact of the recent mineral exploitation in Papua New Guinea:

> ... landowners are acting out an ideology of landownership which has its own history, and which colours the definition of compensation in

> particular ways. [People] may fail to see that when landowners become engaged in a relationship of compensation with some external agency, their status as landowners (and their consequent role within this relationship) is not a simple and straightforward fact of life. For there is a sense in which Papua New Guineans have only *become* landowners over the course of the last 10 years … the question of whether 'clans' exist as 'landowners' in the fabric of national identity is the question of how 'clans' *have actually become* groups of landowners claiming compensation from development of their own resources. [Filer 1997a:162, 168]

Consistent with this insight of Filer's, the approach of the authors in this volume goes beyond the assumption that indigenous landowners are also, like their Western counterparts, a species of *Homo economicus*, seeking to derive maximum material advantage from the presence of mining projects (though some of the chapters in Toft 1997 attempt to question this *a priori* assumption). Of course it is true, but given that most local landowners do not understand the complex financial arrangements by which mining revenues reach them, they are apt to interpret and anticipate such benefits in more qualitative terms, which Filer's work has highlighted (Filer 1997a). We seek to describe what 'advantage' and 'material' mean in the local vernacular, and how these things commonly have a magical and/or mythopoietic dimension quite different from Western understandings.

The devil and mineral extraction

Another theme we would like to draw attention to is the confluence of apocalyptic imagery and territorial or 'world' transformation, the manner in which capitalism and Christian eschatology are appropriated within the local mythopoieia of mineral deposits. Stürzenhofecker (1994:27) has thus written that Duna men of the Southern Highlands Province, Papua New Guinea:

> blend received notions regarding powerful spirits with rumours regarding the finding of oil resources, in such a way as to move from the picture of a sacred landscape, whose fertility must be preserved for the future, to a picture of an exploitable landscape available for manipulation by a company … Their peripheral location, coupled with rumours of the centralising potential of company development, have given them an almost apocalyptic vision of what such a form of development could bring to them, regardless of the likely ecological consequences.

Theoretically, we adduce case studies that dramatically underscore some of the main points that Gilles Deleuze and Felix Guattari have made concerning the relation between capitalism and the 'primitive body' (Deleuze and Guattari 1983). What they call the 'primitive machine':

> subdivides the people, but does so on the indivisible earth where the connective, disjunctive and conjunctive relations of each section are inscribed along with the other relations. [Deleuze and Guattari 1983:145]

Taussig's first ethnography can be seen as one of the first attempts to address this scheme of Deleuze and Guattari from the point of view of ethnography. His study of Colombian miners became an account of the fracturing, or decorporealising of this embodied relationality through mining and its attendant capitalist productive regime – the search for minerals became an aggressive act of dismemberment of the earth, a dismemberment which nevertheless is read back onto the territorial logic of indigenous people:

> Their culture, like their work, organically connects soul with hand, and the world of enchanted beings that they create seems as intensely human as the relations that enter into their material products. The new experience of commodity production fragments and challenges that organic interconnection. Yet the meaning of that mode of production and of the contradictions that it now poses is inevitably assimilated into patterns that are preestablished in the group's culture. [Taussig 1980:11]

McIntosh's chapter provides a powerful historical dimension to our understanding of the production of mythic accounts among the Yolngu of north-east Arnhem Land, Northern Territory, showing that Macassan traders demonstrated the process of iron extraction from lateritic deposits in Arnhem Land. This revelation of historic acts of mining became reinscribed into the Dreaming as part of the Yolngu's attempt to gain symbolic control over mineral exploration in Arnhem Land. McIntosh's Macassan blacksmiths in Arnhem Land provide a fine example of Deleuze and Guattari's contentious counter-formulation:

> Speech communities and languages, independently of writing, do not define closed groups of people who understand one another but prima-

> rily determine relations between groups who do not understand one another: if there is language, it is fundamentally between those who do not speak the same tongue. [Deleuze and Guattari 1983:430]

Finally, a major theme that emerges in the consideration of large-scale resource extraction projects is loss: when place and landscape anchor sociality in the profound way they do in New Guinea, the actual physical removal of places through mining is the excision of a fundamental ground of human relationality and what Wagner calls human focality (1998). People such as the Kaluli and Foi of the Southern Highlands Province and the Rauto of West New Britain always speak about the loss of people through death as having an impact on the appearance of the landscape. But we are now confronted with an inside-out version of this: what happens when it is not people who die, but the places themselves? As Kirsch asks in his chapter, what becomes of the 'grounds' of human life under such circumstances? (See Weiner 1998.) Given that in both Australia and Papua New Guinea, specific narrative techniques such as myth, poetry and song are used to make these geographic dimensions of human community visible, a major alteration of the landscape should make itself felt most conspicuously in the verbal behaviour of humans who constitute it within the register of language. Thus, as Andrew Lattas has suggested in a comment on Kirsch's chapter, one cannot speak of 'a sense of place' without appealing to the invocation of loss and absence through which place intrudes on human consciousness. Stories and poems about places are characteristically about the actions of ancestors who are no longer present in human community but who left marks of their activities on the landscape. Poetic songs in Kaluli and Foi speak of the poignancy of places left abandoned through the death of people. The positive tracery of marks and signs visible in the landscape therefore always speaks of an absence, without which the sense of transformation through movement would not be possible.

We can call this a cosmology of emplacement as long as we accept that this mode of emplacing is a temporality-producing procedure as well. Moreover, the 'production of time' through space continues in contemporary Aboriginal and Papua New Guinean narrative techniques and strategies. Throughout Australia and Papua New Guinea, ancestral events were also historical events. Thus, the threats of cataclysmic consequences wrought by ancestor beings at the sites of mining and other ground-altering developments at Coronation Hill, Hindmarsh Island, and Porgera are evidence that what we call mythopoiesis continues to have evenamental effects in the contemporary world. The attempts to re-inscribe ancestral efficacy through the medium of mythico-cosmological accounts and predictions are also events in the world, reactions to real-world pressures, imbalances in power and external intrusions, as Merlan has suggested (1998).

There is thus always a temporal dimension to spatiality and our task becomes that of elucidating the objectifying techniques such as narrative which make the elicitation of this movement possible. But as Kirsch reminds us, 'these narrative conventions are only powerful and compelling because they have effects in the material world, and when that world alters profoundly, the power of speech itself may become attenuated'. Thus, it is not just in mythological terms that being in place allows one to speak from a position, as it were – the intimate relation between place and its mythopoietic manifestations demonstrates the manner 'in which landownership confers "voice" – the right to speak and the ability to influence the flow of benefits from the land its resources' (Ballard 1997:48).

But this confluence of land and voice that we are trying to characterise works in both directions. In Australia and Papua New Guinea, we witness as Filer has observed 'the incorporation of the local community through the adoption or imposition of bureaucratic management methods …' (quoted in Sagir, this vol.). The legal mechanism of incorporation is not a neutral organisational and administrative template: like the mythologies that indigenous people bring to bear in the face of their own explanatory challenges, it is a systematisation of core components of personhood, autonomy, consensus and will in Western culture. As such, the language of incorporation, so widely adopted by indigenous communities in their confrontation with governments and companies over control of resources and land, has been the most effective agent of their own sociopolitical transformation. Whether a truly indigenous form of the corporation, something we took for granted in Radcliffe-Brown's heyday but which seems a far more perilous a proposition now, can emerge from this conjunctural arena is only one of a number of questions that an anthropology of indigenous communities in Australia and Papua New Guinea will have to closely attend to now.

2
THE IRON FURNACE OF BIRRINYDJI

Ian McIntosh

Scattered throughout north-east Arnhem Land's communities and former Yolngu (Aboriginal) living areas, are replicas of a ship's mast. At one level they represent the 'chimney' of the iron furnace of the Dreaming figure Birrinydji. At another level they stand for Birrinydji himself, and 'remembership' of a time long ago when Aborigines enjoyed the wealth and privilege that now mostly whites do. In a Warramiri clan perspective, Birrinydji is the king, the boat captain and blacksmith. He is at once the minerals of the land and the land itself, and Yolngu are born in his image and ceremonially enact his will. Yet Birrinydji represents the technology and power possessed by non-Aborigines, but not Aborigines. The perception is that following the days when Indonesian fishermen from Macassar frequented the northern coast, Balanda (non-Aborigines) became wealthy at Aboriginal expense by exploiting the resources of the Aboriginal domain. In an analysis of Aboriginal oral history and of the spirit-being Birrinydji, I will in this chapter attempt to account for widely varying attitudes of Yolngu people towards mining and exploration in northern Australia.

Contemporary Arnhem Land

The geographic focus of this chapter is the Aboriginal community of Galiwin'ku (Elcho Island) and also the homeland of Dholtji in Australia's Northern Territory, both some 500 kilometres east of the regional capital of Darwin. The largest settlement in north-east Arnhem Land, Galiwin'ku was established in 1942 by the Methodist Overseas Mission. The community is home to approximately 1500 out of a total Yolngu population of 5000. The traditional Aboriginal owners of the island share their homeland with eight other closely related clan groups, whose country lies in the immediate vicinity of the settlement. Dholtji is a small outstation to the north-east of Galiwin'ku and has a population of perhaps ten people, but only during the dry season.

The people of north-east Arnhem Land are known as Yolngu, but they have also been referred to in the literature as Murngin (Warner 1969) and Wulamba (Berndt and Berndt 1954). Dualism is the defining feature of the Yolngu universe. Each person is born into a patrimoiety, named Dhuwa or Yirritja, and a patrilineal landowning clan, which again is either Dhuwa or Yirritja. Each member possesses rights to access and use certain lands through their father, who gives a person their clan identity, but they also have interests in the land of their mother and their mother's mother. In this chapter attention is focussed on the Yirritja moiety Warramiri clan, and in particular the views of the former clan leader, the late David Burrumarra, MBE.

The traditional owners of Galiwin'ku and Dholtji are land owners in the eyes of the Australian law. Yolngu property rights came under the national spotlight in the 1960s, when Aborigines living at nearby Yirrkala took their case against the mining company Nabalco to the Supreme Court to determine who owned the land. As is well documented (Williams 1986) , the judgement was that while Aborigines belonged to the land, it did not belong to them. The court did not recognise community or group land interests and the decision was that Aboriginal property rights had been wiped out with the assertion of sovereignty by the British in 1788.[1] The decision of the court prompted the establishment of a commission of inquiry into granting land rights to Aborigines in the Northern Territory. This led to the enactment of the *Aboriginal Land Rights Act* (NT) 1976, and the formation of land councils to pursue land claims on behalf of traditional Aboriginal land owners and to act as liaison with regard to development proposals. Aborigines, who make up 25 per cent of the Northern Territory's population, now own over 50 per cent of its land mass. The form of title is inalienable or Aboriginal freehold. Aborigines hold title not just for themselves but for future generations. The land cannot be sold or given away.

Introducing Birrinydji

For at least 200 years, fishermen from southern Sulawesi, the Macassans,[2] made annual voyages to the Arnhem Land coast in search of the exotic delicacy, trepang, which they would sell to the Chinese. Yirritja moiety Warramiri clan members once had a pivotal role mediating relations with Macassans. In a process that was to be echoed elsewhere in Australia by European colonists, Macassans bestowed the title Rajah (king) on certain Aboriginal elders who would act as their brokers. In turn, these leaders would regulate trade in highly desired foreign goods with inland Aboriginal tribes (Thomson 1949). The last Rajah of Dholtji and Melville Bay was Ganimbirrngu (Macknight 1979) who was the father of David Burrumarra, the immediate past Warramiri clan leader from Elcho Island.[3] Dholtji was the place where Matthew Flinders encountered the Macassan fleet in 1803[4] and where great festivities are described as taking place on shore as up to sixty praus and a thousand men reconnoitred prior to their return to Sulawesi (Berndt and Berndt 1954). Following the departure of the Macassans in 1907 and the establishment of missions in north-east Arnhem Land in the 1920s, Dholtji became all but deserted.

In 1988, eighty years after the end of the Macassan era, David Burrumarra promoted the view amongst those who would listen that mining on his land at Dholtji would help restore wealth and status to Aboriginal people – a wealth and status that had been usurped firstly by these itinerant voyagers from Sulawesi and then later by white settlers. For the aging Aboriginal leader there were no doubts. Mining was a part of Aboriginal history and it was an avenue to the good life. Since time immemorial, coastal haematite outcrops had been transformed into iron-bladed tools by Yolngu working under the guidance of the Dreaming figure Birrinydji. Burrumarra sought a return to this 'golden era'.

Under provisions of the *Aboriginal Land Rights Act* (Northern Territory) 1976, Aborigines must respond to requests for a meeting with potential developers not less than every five years. In 1988, as with previous negotiations, Burrumarra's position remained the same, though he lacked support from many members of his clan. The destruction of sacred sites in the vicinity of the Nabalco bauxite mine at Gove was still fresh in people's minds, even twenty years on. While close family did not question Burrumarra's authority as the spokesperson for the clan or Birrinydji, some Yolngu saw him as attempting to sell off the country to Balanda for his own personal gain. Others saw his views on the past as obscure and anachronistic and they objected to mining exploration, even while acknowledging that the land in question was sacred to the memory of a 'timeless' partnership that was once deemed to have existed between Aborigines and non-Aborigines through Birrinydji. As Burrumarra said:

> Birrinydji had the mind of a Balanda ... but his law is for all, not just for brown and white, but black as well, and all the people of the world ... Birrinydji was the king just as my father was king. My father was also a servant ... When he looks in the mirror he sees Birrinydji, but also the whale and the octopus.[5] The Warramiri honour all three. [McIntosh 1996]

Warramiri oral history tells a tale of first contact and a history of race relations that is at odds with contemporary historical accounts of the Macassan era. This chapter revolves around discussions with Burrumarra and other members of the Warramiri clan and Yirritja moiety on the history of iron-making in Aboriginal Australia. It examines the relevance of this history in the way that Aborigines in north-east Arnhem Land are responding to requests for access to their country by mining companies.

Burrumarra's dream

In the weeks following a 'yes' decision to mining at a Northern Land Council meeting held at Galiwin'ku to discuss the possibility of exploration in the vicinity of his outstation, Burrumarra had a dream which sparked considerable discussion. In the dream Burrumarra's younger brothers were clearing land for an airstrip at Dholtji,[6] at this ceremonial centre for Birrinydji. In complying with Burrumarra's wish, the brothers were planning to make Dholtji the large settlement it had been both during and prior to the Macassan era.

The brothers had nearly finished the airstrip when their bulldozer was halted by an obstacle. It was a huge gold nugget. 'This must belong to Birrinydji' they thought, and went off to get their older brother. The Warramiri leader stared at the find and understood that the wealth of the white men could be theirs once again, and he reflected on the past. Birrinydji was the rich minerals of the earth, the transformed haematite, the source of the technology that made foreigners wealthy and allowed them to dominate Aborigines. He bent down to pick up the prize and, as he lifted it and held it in his arms, Birrinydji, Burrumarra's Aboriginality and his Dreaming went into the ground and out of his life. He had the wealth of the Balanda and that was all that he had. To savage the earth for its 'spirit', as in mining, was to lose one's identity and become like the Balanda. By resisting the temptation one would maintain one's Aboriginality but also one's poverty.

Theoretical overview

The historian Campbell Macknight has called for an analysis of the way the memory of the often turbulent indigenous experience with Macassans has been transformed by Aborigines over time and is relevant in contemporary politics (Macknight 1972:317; 1986:72). Keen likewise has suggested a need to 'trace trajectories of transformation in relations, powers, trends, events, and the forms into which people try to shape their world' (Keen 1994:296). In this chapter I ask, 'What is the relation between giving an account of the past and realising a future?' What happens when 'timeless' truths of the Dreaming are in conflict with prevailing or available data?

Bloch (1977) referred to a distinction between systems by which we know the world in a practical sense and systems by which we accommodate history and keep the law. This was equated with Marx's distinction between ideology and knowledge. For the Warramiri there is knowledge of the Macassan past that pertains to everyday communication, (recently introduced to Arnhem Land schools is a course on Macassan history) and there is knowledge of Macassans applicable to ritual communication amongst elders in a ceremonial setting (for example, Birrinydji comes from the land of the Warramiri and drew all outsiders to Arnhem Land by the strength of his *marr* or desire). 'Outside' (or public) and 'inside' (sacred or esoteric) are apt labels for the distinction. According to Bloch, anthropological analysis must take into account the changing meaning of the past in the present. Answers to questions such as: what is known about Macassans, what is possible to know, and who has the right to speak on this subject, reflect the interaction of emic and etic perspectives. The ancestral being Birrinydji is constituted in the ever-changing relations between Aborigines and others, and Yolngu decision making with regard to mining reflects the ever-changing understanding of this Dreaming.

In the 1960s, Warramiri Yolngu worked with historians charting Macassan sites along the coast. The subject of Birrinydji was not raised, and one of the reasons was that 'inside' truths and available or prevailing 'outside' data did not address the same questions, and some Yolngu wanted to keep it that way.[7] For Burrumarra, only the Birrinydji narrative was seen to provide answers to questions such as: What brought the Macassans to Arnhem Land? Why do some Aboriginal Dreamings refer to ancestral figures in the image of the Macassan? Why do some Yolngu have a ceremony for iron and not others? Burrumarra did not know that in the 1700s, Sama Bajau, or sea gypsies, were scouting for the Macassans – seeking out new areas for exchange and exploitation. Similarly Yolngu were also not privy to the political events in Southern Sulawesi in the 17th century, when the Dutch, in league with the Bugis, took Macassar by force and changed the nature of sea trade in the Indonesian archipelago forever. What they did know was that Macassans were white and rich and Aborigines black and poor, and Aborigines had to work for the visitors to get what they wanted in the way of trade goods.

Easy access to iron since the advent of the mission at Galiwin'ku in 1942 has resulted in a change of status for Birrinydji. As a foundational story for clans such as the Warramiri, can it simply vanish into obscurity? Following Sahlins (1981), existing interpretations are seen to be put at risk by Burrumarra. Sahlins says that just as history is culturally ordered, so too are cultural schemes historical. We take risks with our understandings, and as a result, culture is historically altered in action.

Everything that exists has an 'inside' equivalent, which may or may not have been revealed at a particular point of time, allowing the 'inside' to always appear to be unfolding before one's eyes. This was central to Burrumarra's understanding of the need for mining. Aboriginal history was suggestive of a potential that was yet to be realised, and he was testing the waters. Burrumarra believed the time was right for mining, but was uncertain of the extent of support for his interpretations of the significance of Birrinydji. Burrumarra's appeals to the Dreaming were based on a desire for an end to the poverty which for so long positioned Aborigines on the margins of non-Aboriginal society. The question is how, in giving his account of Birrinydji, did Burrumarra hope to make the future coincide with his vision of the past? And what becomes of Birrinydji in the process?

The history of race relations in Arnhem Land has been such that Yolngu feel great resentment towards the Balanda. Yet throughout the Yolngu territories, Aborigines differ in their opinion of the Birrinydji legacy. While Dhalwangu and Gumatj clan members treasure their detailed knowledge of his songs and ceremonies, and carefully maintain sites in the landscape, they are not able or willing to speak about the related narrative. The Birrinydji Dreaming has become all but a mythless rite. The Warramiri, on the other hand, have elaborate stories, but they are cautious about revealing them. 'Whites' might come to believe that the Dreaming condones their discriminatory practices, Burrumarra's brother Liwukang declared.

Iron use in north-east Arnhem Land

There are no records of Aboriginal iron-making in precolonial Australia and similarly only scant references to the mythological significance of iron to Aboriginal populations in the early years of European settlement. Reynolds (1990:48) shows that some groups had terminology for the various products of the blacksmith's trade and had worked iron weapons while the metal was hot. He also says that there was an early and widespread adoption of the use of iron by Aborigines following contact with Macassans, though Warner suggests that the use of metal by Yolngu may have preceded the arrival of Indonesians. Wooden planks with nails attached would have been continually floating onto the coast from the north and north-west and might have been extracted and used as fish hooks.[8]

A major outcome of the Macassan period was an appreciation of iron's unique qualities and it became a highly prized item of trade (Warner 1969:450; Macknight 1972:305). Particularly valued items were the tomahawk and knife, the detachable harpoon head, shovel-nose metal spear and the small metal bowl used in long wooden smoking pipes. According to Thomson (1957:31), the Yolngu were good at working in metal, making fine spear heads by beating out cold odds and ends of scrap metal and rigging screws, but there is no suggestion that such techniques were passed on as a result of contact with Indonesians. He writes:

> [Aborigines] … made fish hooks and even knives from the nails and other fragments of iron that they salvaged from planks of driftwood, or iron from water tanks and trepang boilers of wrecked ships.
>
> One of their most enterprising ventures in quest of iron occurred in Melville Bay where the Royal Air Force had anchored drums to serve

> as mooring buoys for … flying boats which had to refuel there. The [Aborigines], alive to the value of this iron within their reach, cut the drums adrift, beached them and cut them up into sheets of iron to be beaten into spears. [ibid.]

Just as there are no records of Aborigines mining and smelting ironstone, there are no records of Macassans making metal tools on the Australian coast, although they definitely had an interest in prospecting. Earl was stationed at Port Essington in the early 1800s and conducted a regular trade with the trepangers. He says that in the vicinity of Elcho Island and Arnhem Bay:

> This part of the coast is apparently the termination of a granite range, and is said by the Macassars to abound in minerals, among which they mention tin, but … appears to me to be antimony-ore which will yield perhaps two-thirds of its weight in metal. [Earl 1842:141]

Searcy traversed the coastline in the late 1800s and he also mentions this interest by Macassans. At a trepang smoke house in Melville Bay he:

> found specimens of quartz and ironstone, in one of which a speck of gold could be distinctly seen … There was also a stack of manganese, which commodity for some reason the Malays [Macassans] took to Macassar. [Macknight 1976:44]

Elcho Island was also a source of 'red pigment' for the Macassans, but their interest in this material is unclear.[9] According to Burrumarra it could have been one of two types. Macknight[10] says Burrumarra had suggested this red clay was perhaps that which Aborigines collect from nearby Howard Island. Called Miku, it is dug out from a cleared area known as Gulpulu, burnt on the fire and then applied to the body in preparation for ceremonies. The alternative was the Dhuwa moiety red rock *rratjpa*, which is the source material associated with Yirritja moiety iron production. This red laterite is found in abundance in the cliffs at Galiwin'ku, Elcho Island, adjoining an old Macassan trepanging site. It is haematite (70 per cent iron) (Dana 1949:484), a variety of iron ore used not only in the production of steel, but also commercially in the production of red paint.

Could Arnhem Land have been a source of raw material for local and overseas iron production? Macknight's detailed study of the Macassan trepang industry has shown that the visitors usually spent not more than a week or two in any location, but there are various recordings of visitors having to spend extended periods after being shipwrecked or failing to catch the trade winds in time for the return journey to Sulawesi. In normal circumstances however, as Macknight (1972:309) infers, it would be 'most unusual … during a voyage of this character'. Yet one could certainly imagine the types of situations where iron-making might have become a necessity say, for example, if the anchor was lost at sea or if nails were required to repair the praus and there were no other craft in the vicinity to lend assistance.

The process of iron manufacture does not require elaborate machinery. Any place where raw materials are available will suffice. In fact, techniques which might have been practiced in Arnhem Land in the past are still carried on throughout eastern Indonesia today. As Reid says:

> The characteristic Southeast Asian bellows – two vertical tubes with pistons lined with chicken-feathers, pumped by an apprentice sitting above them – is everywhere still in use. The remaining equipment is very basic – anvil, various hammers, a cutting wedge, tongs, scraper, and a bamboo full of water for tempering the steel. [Reid 1983:19]

Harrison and O'Connor describe the process in more detail. They write:

> A furnace simply consisted of a sort of circular … pit, three or four feet in diameter, dug in compact earth … The pit was connected with a circular hole above … through which the smelters subsequently added supplies of charcoal … After igniting the charcoal they closed the mouth of the pit by means of earth to keep back the heat and … to melt the ore. They then allowed the molten metal to flow out by tapping the lower part of the furnace and the slag was separated. [Harrison and O'Connor 1969:313]

Macassans may well have made iron on the Arnhem Land coast and Aborigines of Burrumarra's clan either witnessed this or participated in its production. The many Aborigines who travelled to Macassar during the 200 years of the trepang trade would also undoubtedly have come across the industry. For instance Burrumarra, drawing his knowledge of the iron-making process from oral history and the songs of Birrinydji says:

> Birrinydji used the 'red rock' from the beach, not bauxite, that's only for Gunapipi.[11] [The red rock is] called *rratjpa* … and comes from

Djang'kawu [a Dhuwa moiety ancestral figure]. At Cape Wilberforce they call it *mirrki*, red sand of the sunrise. Red rock is intelligence for all mankind, the source of wealth and power of Balanda and Yolngu – from it comes all the technology – axes, knives and hammers.[12]

While the 'red rock' is linked to Dhuwa moiety ancestral and totemic themes, Birrinydji is of the Yirritja moiety. When the raw material is transformed by fire it enters a new domain separate from its former associations. Thus while in the Dhuwa moiety there are countless myths about *rratjpa*, none is associated with iron production.

Details of iron-making are recorded not only in oral history and Yirritja mythology but also in art, song and in the personal names of Yirritja Yolngu. In the Warramiri clan alone, more than 20 per cent of the registered first names of clan members are drawn from the Birrinydji theme; in other coastal Yirritja clan groups, the figure is less but averages over 10 per cent. Cawte (1993:44) details one song about knife manufacture in his book on the Warramiri. In this and related songs, the singer identifies himself with the technology of Birrinydji as well as the manufacturing process. It goes thus:

Ngayum djangu latimi
I am the blade

Ngayum djangu djidami
I am the handle

Ngayum djangu wambalmi
I am the long knife

Ngayum djangu butumi
I am the wood for a handle

Ngayum djangu rrawarra
I am the steel template.

Berndt (1949:221) also refers to Warramiri ceremonies associated with iron-making. He says the Gwolwunbuma, Lil'garun, Mararaguma, or Jandyaralguma, are connected with the shovel-nosed iron-bladed spear, the knife and the axe. Elkin likewise, in one of his Warramiri recordings from north-east Arnhem Land says that:

'The song of the anvil' describes vividly the darting of sparks and the 'cry' or resounding noise when the heated iron is struck with the hammer. [Elkin 1953:91]

Burrumarra says that all of these songs are from Birrinydji and are related generally to the idea of white and black men working together, for the legacy of this Dreaming figure is centred on the concept of a partnership between peoples under the one law.

The legacy of the dreaming

In the 1940s, the Berndts found pottery fragments at a Macassan trepanging site at Port Bradshaw in north-east Arnhem Land. Aboriginal informants stated that such pots had been made by them from local ant hill in the pre-Macassan past. Songs recorded by the Berndts talk of this production.[13] According to Aboriginal oral history, pottery making was a woman's job and it was the legacy of Birrinydji's wife, Bayini, to her historical female counterparts, both Aboriginal and non-Aboriginal. The same was the case for both cloth manufacture and weaving. Rice production, likewise, was carried out by Aboriginal women in Gumatj, Dhalwangu and Warramiri clan territories (Berndt and Berndt 1954:37; Mountford 1956-64:295). Informants can still point out the old paddy fields which today are usually associated with large fresh water billabongs. The rice has turned into *rakay*, the water chestnut, another significant Yirritja moiety totem. But there was and remains considerable variation in Aboriginal accounts of this 'pre-Macassan' period, because they entail a paradox. Stories stress Aboriginal wealth and self-sufficiency but also their subsequent loss in relation to outsiders. As the Berndts suggest, Aborigines did not desire to imitate the Bayini, preferring their own way of life and while the two groups coexisted, they did not seem to be willing to learn from each other. The Bayini, they say, kept the secret of weaving to themselves (Berndt and Berndt 1954:38).

In contrast to the occupations and technologies of the female ancestral being Bayini, the mining and smelting of iron ore was work for men, and Birrinydji, the 'man of iron', instructed Aborigines in this trade. According to Yolngu, references to the making of iron are thus wrongly attributed to the Macassan era. They say it was in the 'pre-Macassan' or Murrnginy[14] period, the 'golden age' of Birrinydji. According to Burrumarra:

> Macassans had Birrinydji in common with Arnhem Land but the spirit of Birrinydji is Dholtji. All things came to the Warramiri from Birrinydji and then to other clans.

The Birrinydji dreaming

Just as a totem represents the outward form of a Dreaming being, a Macassan *bunggawa* (boat captain) by the name of Luki appears to provide a visual image of what Birrinydji is like. Otherwise, Birrinydji is indistinguishable from other Dreaming figures. Sacred *rangga* that represent his legacy are the basis of extensive clan alliances within the Yirritja moiety or half of Yolngu society. Numerous totemic species such as the swordfish, angel fish (with fins like the sails of a boat) or even the Gawukal, a bird with a tail resembling an axe, owe their form to his intervention, and Birrinydji is associated with specific tracts of country belonging to the Warramiri, Dhalwangu and Gumatj clans.

All that is known of the Birrinydji Dreaming has been passed down to the present through many hands and interpretive processes. The Warramiri leader Bukulatjpi, for instance, is credited with 'doing the thinking' and uncovering the 'truth' about Birrinydji and the Macassans. Living in the mid 1800s, he was the first to do Birrinydji's dance and pass on its meaning to others. His descendants Yamaliny, Lela, Bambung and Ganimbirrngu were all 'frontmen' or brokers for the Macassans.[15] They all had Macassan names, were followers of this new law and were staunch defenders of their lands against unwelcome intruders. Ganimbirrngu's son, David Burrumarra, was born ten years after the trepang era ended. For Burrumarra, the Birrinydji Dreaming was to become a strategy for survival. In present-day understandings, Burrumarra's legacy is Birrinydji's. Through his interpretations, this Dreaming narrative became a foundation for Warramiri belief in Christianity, the moral basis of land and sea claims, and the rationale for a treaty or pact of reconciliation between Aborigines and non-Aborigines (see McIntosh 1995).

In many parts of Australia the first settlers were deemed to be ancestors returned from the dead, their light skin colour being evidence of the bodily decay that takes place in the weeks following a death. Bukulatjpi, living on remote Cape Wilberforce in the mid-1800s, understood that Macassans were not Yolngu or in any way divine. Rather, their inordinate material wealth and willingness to share this with Aborigines in exchange for labour was having a major impact on Warramiri lifeways. Their possession of modern technology was equated with the visitors being recipients of

Birrinydji's inheritance. Bukulatjpi reasoned that something had gone wrong at the 'beginning of time' for this was an Aboriginal Dreaming and its inheritance was the right of clans such as the Warramiri, as long as the Birrinydji ritual was performed. But it was the Macassans that performed their dances on the beaches of Arnhem Land. Yolngu had long 'forgotten' them.

Bukulatjpi viewed Birrinydji as a Dreaming figure that controlled the seasonal movement of the Macassan trading fleet and also the winds that would bring the visitors onto the coast each November. Birrinydji also provided these fishermen with skills to fashion swords and knives and pottery with magical precision, and grow rice and other plants foods from Warramiri billabongs. But Birrinydji did not act alone. Bukulatjpi understood that he was answerable to a higher Dreaming authority – Walitha'walitha[16] or Allah.

Birrinydji and his wife Bayini were ancestors of Yirritja Yolngu – creational figures that emerged from the Australian mainland at a point beyond memory. Birrinydji ordained that certain non-Aborigines (that is, the entire non-Aboriginal population) would come to Arnhem Land to make the land and the people strong. Bayini's children would introduce the Yolngu to the new world – to 'bring them up to date'. First there were black whale hunters from the mythical islands of Badu to the north-east of Galiwin'ku; then golden-brown workers for Bayini, also known by the term 'Bayini'; then light brown Macassans from the north-west; and finally white Japanese pearlers in the 1920s and European colonists. This colour change in the visitors from black to white corresponds with a change in attitude towards reciprocity in dealings with Aborigines, and therefore knowledge or ignorance of Birrinydji's law.

Aborigines and whale hunters were united in Walitha'walitha through the whale, an outside symbol for Birrinydji. Together they upheld the law of the sea. The alliance between the two was such that the souls of the Aboriginal dead from the Yirritja moiety went on the backs of whales to the land of the dead, guided by these hunters. The Bayini, on the other hand, after introducing the laws of Birrinydji to the Yolngu, kept the secrets of iron-making and weaving to themselves when they departed. Certain Macassan leaders recognised the law of Birrinydji and respected Aboriginal sovereignty, but most did not and there was great disparity in wealth between the nations, and trade relations were far from savoury in the latter period of the industry. Finally, Japanese and Europeans totally ignored Aboriginal rights and there was no reciprocity in the relationships.

By far the most significant of the waves of contact was the golden-skinned Bayini, the bringer of Birrinydji's laws to Aborigines. In a perspective from the Dreaming, the Warramiri homeland of Dholtji became a centre for iron manufacture, boat building, and rice, clothing and pottery production. When united, Aborigines and the Bayini prospered, but over time their fortunes soured. Warramiri Aborigines desired only good, but 'bad came too'. A 'fire came to the Yolngu' and 'there was great bitterness between white and black', Burrumarra said. The spirit of the dead or Wurramu took over Yolngu lives. Birrinydji wanted to bring more non-Aboriginal people to Arnhem Land but Walitha'walitha sent the newcomers and Birrinydji away, for Walitha'walitha could see how the Yolngu were suffering. So the Bayini left and their parting words to the Aborigines, 'From now on you must look after yourself', were the signal for the beginning of an era of impoverishment. The visitors had not finished the work of teaching. 'Birrinydji did not want to stay in Australia,' Burrumarra surmised, 'but he left the Wurramu and Walitha'walitha here.' Birrinydji and Bayini's legacies are the ongoing and unsavoury consequences of contact, and ideas of the good life and salvation in Allah.

The Yolngu did not cope well with the new world that Birrinydji had offered to them, for *yatjkurr* or 'unholy' ideas ruled their behaviour. This is one interpretation that the Warramiri posit as the reason for joining the mission at Galiwin'ku in 1942. They looked to Balanda missionaries for another chance at life in the new world. Promised was a return to the wealth that was lost at the 'beginning of time', and all that the Yolngu had to do was follow Church law. But it soon became apparent that European and Tongan missionaries had little to share with Aborigines except ideas of a paradise to come in the after-life. According to the Warramiri, it was a fundamental miscarriage of justice. Balanda had the proceeds of Birrinydji, but they were completely ignorant of his law. As these were people who shared a common spirit through the Dreaming, there should be equity in the relationship.

This constructed past can be seen from another angle. In the Dhuwa moiety is the Djang'kawu narrative, which is relevant to the entire Yolngu population, for the mother of a Yirritja Yolngu is always Dhuwa. At one level, it deals with the loss of power of women to men at the 'beginning of time.' Women alone once safeguarded the sacred *rangga* and performed the necessary rites to 'uphold the universe', but one day, the *rangga* were stolen by men. From that point onwards, women had to look away when men performed the sacred ceremony associated with Djang'kawu, for they did it so well. In Burrumarra's (and Bukulatjpi's) narrative, Aborigines were once powerful, but now 'whites' keep that power to themselves. But just as man cannot exist without women, non-Aborigines need Aborigines. Their wealth and influence comes from Aboriginal land and an Aboriginal Dreaming, just as a man's body, his *rumbal* or truth, comes from the woman.

The dilemma of mining

Stanner (1984) says that for Aborigines the present is determined by the past. There is a complete subordination of history to the ideology of the Dreaming. Burrumarra would have agreed. To follow the law and realise a pre-ordained future, mining must occur, but only on Aboriginal terms. But then there is the legacy of other creational figures such as Lany'tjun, the founder of the Yirritja moiety. Each Yirritja Yolngu clan was ascribed certain territories by Lany'tjun to care for, at the 'beginning of time'. So there is a tension here, and this was evident in Burrumarra's dream. For many Yolngu, mining is not an option, and the very idea of non-Aboriginal companies drilling on Yolngu land evokes a deep bitterness, especially in the Gove area. Some see Birrinydji, and consequently their Aboriginality, as being vulnerable to such desecration. For example when satellite mining exploration photos were taken in the Gapuwiyak area in the late 1980s, without Aboriginal consent, Birrinydji was seen in the shadows running from the camera.[17] This fear of mining by Yolngu people is well documented. In relation to a painting of Birrinydji with the metal tools of his trade, Cawte says:

> Warramiri contemplating Birrinydji are supposed to ponder why their 'iron age' was lost ... Does an iron age destroy itself because mining violates the earth? [Cawte 1993:68]

While Burrumarra linked the extraction of bauxite and the production of alumina at the Nabalco plant at Gove with Birrinydji's iron-making, he was against the mine from the outset because of a failure on the part of developers to consult with Yolngu. Yet as far back as the 1950s, Burrumarra and other Yolngu leaders had tried to negotiate mining deals for the extraction of bauxite from the Wessel Islands. In the plan that was envisaged,[18] Aborigines would have a controlling interest in the project and there was a guarantee that no sacred sites would be interfered with. The negotiations entered into with missionaries and others predated by over twenty years the *Aboriginal Land Rights Act* and the powers it grants Aborigines.

The Warramiri leader was also involved in discussions leading to the establishment of land councils in the Northern Territory. For him, such representative bodies are, in part, the realisation of Birrinydji's plan for Yolngu. Here was a body funded by non-Aborigines whose charter was to act in ways conducive to Aboriginal interests, hinting at the time when both whites and blacks were followers of the one law. The implication of course is that Birrinydji only exists so long as there are divisions in material well-being between cultural groups. From Burrumarra's perspective, mining should be allowed on Aboriginal land but only as long as Balanda respect Yolngu wishes, listen to the land owners and share equally in all proceeds. Many Yolngu make their decision about mining on the basis of this Dreaming precept.

In the Northern Territory, the *Aboriginal Land Rights Act* (NT) gives Aborigines the power of veto over development. In the case of mining exploration at Dholtji in 1988, while Burrumarra's dream did not change his feelings on pushing ahead with the project, his family marked off so much of the exploration zone as sacred and 'no go' areas that it was not feasible for the company to proceed. In 1996, two years after Burrumarra's death, the present day leader of the Warramiri, Wulukang, said 'no' to mining even though a majority of Burrumarra's family were now in favour of opening up their land. The country was too sacred.

For a range of reasons, many connected to his own family history, Burrumarra's answer for his own country was always 'yes', though the history of race relations in north-east Arnhem Land worked against his wish. The people, as a whole, usually say 'no'. Past and bitter experience with mining companies, as well as damage to sacred sites, environmental pollution, and the social impacts of royalty payments, all weigh heavily on people's minds when mining decisions are being made. But at the very least, the 'fact' of iron-making on remote Arnhem Land beaches provides an alternative to the sharply contrasting views that to say 'yes' to mining means selling out one's inheritance (as when Birrinydji went into the ground), while saying 'no' results in the maintenance of cultural difference and poverty. For Burrumarra, mining on Aboriginal terms would mean one could be wealthy and simultaneously maintain one's sense of identity and power. To him, this was the legacy of Birrinydji.

Conclusion

The sequence of events in time which is implicit in an 'inside' reading of the Macassan past can be interpreted as providing guidelines for attaining a desired future, progressively, in the here and now. The essential ingredients are the rebuilding of ties of reciprocity and respect for land and sea rights. Bakhtin (1981:147) speaks of such a perspective as historical inversion. History is something yet to be achieved. Myths about paradise, a golden or heroic age, or an ancient truth which are in no way a part of the past, can only be realised in the future. Oral traditions relating to Birrinydji represent a potential, a dream of how things should be if the law is followed. In the scenario presented here, Burrumarra brings to the fore a perspective on the past which is considered anachronistic by many. Burrumarra professed certainty, but his dream indicated otherwise. Aborigines throughout the Arnhem Land region no longer believe that the technology of whites comes from an Aboriginal Dreaming and that to enjoy material prosperity, one has to follow the law of Birrinydji and hence go through clans such as the Warramiri to obtain highly prized items of trade. The significance of Birrinydji is being revalued, but to what end? Proclamations by Burrumarra on Birrinydji reflect the group leader's conceptions of Warramiri social identity – where they have come from and where they should be going as a people. Warramiri history is an ideology that links Aborigines, Macassans and other non-Aborigines, and provides a commentary on present day lifestyles and the status of relationships. Rather than a passive device for classifying historical events, the Birrinydji narrative is a program for orienting social, political, ritual, and other forms of historical action (cf. Turner 1988:23). For Burrumarra, the proclamation of the *Land Rights Act* and the growing influence and affluence of Aborigines in Australia has created the circumstances whereby mining can and should occur. That the two go together – a social environment of reconciliation and the willingness of mining companies to enter into negotiations with

Yolngu – is seen to be a part of Birrinydji's plan. It is the realisation of the Dreaming in the here and now. History will have achieved its potential.

Acknowledgments

This article is a revised version of one published earlier in *Aboriginal History*, Volume 21, 1997. The article is included in this volume with the kind permission of that journal's publisher, Aboriginal History Inc., Canberra, Australia.

Notes

1. This decision was overturned in the Mabo No. 2 High Court case in 1992.
2. This is an expression used by Aborigines to refer to all foreign trepang (beche-de-mer) fishermen. While primarily from Macassar, boat crews were drawn from many of the islands of eastern Indonesia.
3. McIntosh 1996a. Rather than bestowing the gorget upon Aborigines, Macassans offered Burrumarra's father a ceremonial mast, flag and sword.
4. Flinders 1814. Flinders named the surrounding islands after 'the venerable gentlemen' of the English East India Company.
5. These are both Warramiri totems.
6. *Dholtji* is a loan word, from the Portuguese via Macassans meaning 'the gift'. Dholtji is the gift of Birrinydji to the Warramiri people.
7. Burrumarra, pers. comm. 1988. No Birrinydji sites were visited despite the fact that some coincide with trepang sites.
8. Warner (1969:450). Macknight (1972:304-305) says there is little evidence to support this view.
9. See account of Daeng Sarro in Macknight (1979:183).
10. Macknight fieldwork notes 1967.
11. Sacred Dhuwa moiety ceremony.
12. Burrumarra, pers. comm. 1988.
13. Berndt and Berndt (1947:136). Subsequent analysis of the pottery fragments has not revealed their age, but has indicated that their origin was Indonesia, in all probability the Kai Islands of Maluku. See Key 1969.
14. Murrnginy or Murngin, was the word that Warner 1969 used to refer to Yolngu. It means 'the iron age of Birrinydji' according to Burrumarra.
15. Ganimbirrngu was referred to as the last Rajah of Melville Bay.
16. *La ilaha illa'llah*:There is no god but God.

17. Yepaynga, pers. comm. 1988.

18. While this plan attracted the interest of several companies, it did not proceed.

3

THE MOUNT KARE PYTHON: HULI MYTHS AND GENDERED FANTASIES OF AGENCY

Holly Wardlow

In the Melanesianist literature there is increasing concern about the loss of land to logging and mineral resource extraction (Wood 1995; Stürzenhofecker 1994; Westermark 1997; Filer 1996). Given the importance of the landscape and its features for encoding and memorialising both clan history and intimate affective relationships (Feld and Basso 1996; Weiner 1991; Haley 1996), it is important to understand how Melanesians conceptualise and cope with this loss of land and the colonisation of space more generally. Moreover, as land is relinquished for mineral 'development,' local people increasingly encounter that entity which has come to be known as *kampani* ('the company'), a powerful and sometimes inscrutable form of collective agency that is rapidly changing the economic and social landscape, as well as the physical one. Confronting 'the company' is the modern enactment of 'first contact,' and like the 'first contacts' of the past, the potential for violence is ever-present, and imaginative resources are marshalled on both sides to make sense of disparities in power and competing desires.

Lattas in particular has explored how 'the politics of space' plays itself out in the realm of myth and dream, a realm that he calls an 'alternative geography of hidden places' (1993:66; 1992). According to Lattas, the realm of myth and dream is 'where the white man's structuring and control of space is undermined' (1993:66), and therefore, 'any study of the political space of colonialism must also include the imaginative geography of dreams' (ibid:69). In this chapter I discuss both the 'real' space of Mount Kare as a new site of mineral extraction within the nation-state of Papua New Guinea (PNG), and the indigenously imagined spaces where the production of gold is revealed to be an ancient ancestral process. The 'alternative geography' of Mount Kare unfolds in Huli reformulations of their traditional myths and in more contemporary myth-like narratives that I call 'fantasies of agency.'[1] These 'fantasies of agency' implicitly assert that it is Huli people who ultimately have the true and legitimate power over the wealth to be found at Mount Kare, and they imagine new sorts of social efficacy that make this possible. Importantly, these 'fantasies' are explicitly gendered. While men and women are equally imagining new kinds of individual and relational capacities that would ensure their possession of the gold from Mount Kare, and thus their transformation into '*millionaire man na meri*' (millionaire men and women), the capacities that they imagine for themselves are quite different.

Changing ties to land

Like other Melanesian cultural groups that have been discussed in recent ethnographic work and in this volume, the Huli of Southern Highlands Province in Papua New Guinea have intimate ties to the land on multiple levels (cf. Weiner 1991; Feld and Basso 1996; Haley 1996). Haley asserts for the neighbouring Duna that 'narratives, and the events they describe, are grounded throughout the landscape, as are Duna people's memories' (1996:279), and this is equally true of the Huli. As one moves through the Huli landscape, one learns that even very small areas of land have their own names, and people delight in detailing the series of events that have happened in particular spots as they walk along. Several genres of song, such as those for mourning sung by women and those for courting sung by men, are largely the recitation of specific place names associated with a particular person. Women who attend *dawe anda* (courtship parties) claim that when they hear songs about the places most intimately associated with their lives they become aroused and that the men 'touch' their bodies by singing these names. And as Schieffelin and Feld describe for the Kaluli, the place names detailed in Huli funeral songs invoke specific activities and events associated with the dead (Schieffelin 1976; Feld 1982). In the case of murder, mourning songs are said to have the power to locate the guilty party and to incite men to exact revenge. Emotion and memory is thus 'emplaced,' and places are affectively charged and motivating.

Memories concerning place are also instrumental in asserting claims to land. Huli genealogies are called *dindi malu* (literally, land genealogies), and their correct recitation requires not only knowing who gave birth to whom, but also where individuals were born, where they travelled, where they settled when they married, and where they made gardens, hunted, collected pandanus nuts, and conducted ritual activity. Huli men are known as unusual among Highlands cultural groups for their deep genealogical knowledge, often being able to recite clan histories twenty generations back in time and cognatically across (Ballard 1995). Such knowledge enables men to assert claims to desirable stretches of land. On a more cosmological level, Laurence Goldman, Chris Ballard and others have discussed the Huli sacred geography in which certain features of the local landscape had religious significance, and the Huli designated themselves as responsible for orchestrating the complex, multicultural rituals that were meant to renew the fertility of the land and safeguard the world. Clearly Huli identity is inextricably bound up with memories and knowledge about land.

With the development of mining, *dindi malu* (land genealogies) have become increasingly important as Huli try to assert ownership over land and thus rights to monetary compensation for its use by mining companies (Filer 1990). The Huli recognise the potential claims of all cognatic descendants to the land of a clan founder (Allen 1995), and the Huli kinship system often seems very like a maze: often it is not a question of *whether* or not you are related to a certain clan that is in line to receive monetary compensation, but rather whether you have the knowledge, skill, and perseverance to wind your way back and across through the generations of male and female ancestors to find out *how* you are related to a certain clan. Rights to land also require proof of usage as well as a clan connection, but, in the case of Mount Kare, where high altitude (nearly 3000 metres) has traditionally precluded habitation and gardening, land use has been very easy to assert and very difficult to prove: any number of clans in the area can claim traditional hunting rights. The identification of 'landowners' for Mount Kare has therefore relied heavily on deep mythico-genealogical knowledge and has been a highly contested undertaking. Further complicating the situation is the strong cultural emphasis on geographical decentralisation and the maintenance of multiple residence (Glasse 1968; Goldman 1983). In the past it was an important piece of *mana* (traditional precepts and cultural wisdom) passed on from fathers to sons that brothers should not live together on contiguous areas of land, but should spread out and cultivate different areas. Thus, as Vail points out, the nature of the Huli descent system has, over the generations, resulted in the fact that the land claimants for Mount Kare are highly dispersed (Vail 1995:357). For example, the majority of members of Apai, a clan that is mythically and genealogically associated with Mount Kare and is now claiming compensation rights, now live in Koroba, relatively distant from Mount Kare.

The mining companies' emphasis on attaining 'correct' genealogies as a means of identifying compensation recipients has made for tension not only among men, but between men and women as well. Unlike women in many Highlands cultural groups, Huli women have the right to 'own' land in the same way that men do – they may not alienate the land from the clan, but they can be designated as the person who may choose how it is to be used and to whom it is later given. It is not common for women to own land, but it is acknowledged by men and women to be possible, and increasingly, as more men migrate out of Huli territory for longer periods of time, small parcels of land are given to women by their fathers or brothers. However, the *dindi malu* about particular pieces of land are only articulated in public by men, and men and women both assert that women do not know, and probably do not have the ability to learn, the intricacies of this kind of knowledge (Goldman 1983). One consequence is that women seldom meet with the 'community liaison officers' from the mining companies to assert their claims to land compensation. Men and women often agree that as *tene* (agnates) women have as much right to land 'occupation fee' money as men, and women often vehemently assert that while they do not know and cannot say the *dindi malu*, they 'give birth' to them, since of course it is women who give birth to the individuals who make up genealogies. Moreover, women claim that they are usually responsible for naming children, and that they often deliberately name them after specific places as a reminder of the child's right to that area. Therefore, they conclude, their choice of names also helps to create *dindi malu*. Nevertheless, if women are not physically present at the table collecting 'the company' compensation money, often they simply do not receive any.

While jural ties to land have become ever more salient and fraught with tension, Huli ritual relations to land – in the past so direct and necessary – have been disrupted by technology, and ritual knowledge about the earth that was once very powerful now seems impractical at best and ridiculous or 'satanic' at worst.[2] Moreover, the

Huli have very little understanding of, or ability to participate in, the new modes of relating to the earth – namely, getting an exploration licence and developing a mine. This is not to say that they don't try. During my field work it was believed by members of one clan that another mountain south of Mount Kare would be the next site on which a goldmine would be developed. One of the ancestral figures in the *dindi malu* for this area was Hanamari, a woman who was said to have three heads and to sit on the mountain with her arms folded and her legs crossed (see Fig. 1 in the appendix at the end of this chapter). It was rumoured that a mining company doing a cursory fly-over of the area had seen her through their '*kampas*' (that is, compass, which in Tok Pisin means binoculars), an item of Western technology believed to reveal mythic figures invisible to the naked eye (Clark 1994), and this sighting was an indication that she must be sitting on a huge load of gold. My friends and neighbours were delighted because, unlike the situation at Mount Kare, it was clear which three sub-clans had rights to this mountain from having planted gardens there. Moreover, they thought they were at least one step ahead of the government and 'the company' and that this would give them a tactical advantage: they would form their own landowner company, send delegates to the Chamber of Mining and Petroleum in Port Moresby, the capital city, and apply for the mineral-exploration licence for their own land. They would then simply invite bids from various mining companies and choose the one that made the best offer and had the best record in terms of success and good relations with landowner groups. They even decided it was necessary to appoint three women to the landowner company. There was much joking about this decision since 'no one woman is any better than any other, so how do we decide, and what use are they anyway'. But rumour had it that the inclusion of women was important to mining companies, and so for the purpose of negotiation three women were chosen, of whom I was one.[3] Money was collected, pigs were killed and feasted on, surreptitious meetings were held, and secret lists of clan member names were compiled in the privacy of the men's house and typed up by my field assistant on my mother's forty-year-old manual typewriter.

Little did they know that the exploration licence to their land had already been issued to a mining company which simply had not yet done the exploration. Moreover, to attain a licence one must pay an approximately $2000 application fee, provide an actual plan of exploration, and demonstrate proof of having the necessary technology, know-how, and financing to carry out the exploration. In any case, the Papua New Guinea state owns all mineral resources (cf. Sagir this vol.), only the state can grant leases, and indigenous landowners do not have the automatic right to reject mineral extraction on 'their' land (Nellie James, attorney in charge of mining tenements at the Chamber of Mining and Petroleum, pers. comm.).[4] To the extent that Huli landowners understand the system, they try their best to strategise and leverage themselves into a position of power *vis-à-vis* 'the company'. Unfortunately, they do not have access to information about the system, and even if they did, they do not have the capital or technology to successfully negotiate in it. It is in this context of disempowerment that people conjure up fantasies that reassert their control over, and primary connection to, the land and the wealth to be had there.

Conceptualising 'the company': bankers and landowners

The history of goldmining at Mount Kare has been quite conflict-ridden: there have been conflicts between landowners and 'the company,' conflicts amongst various groups that claim to be landowners, and complex litigation over the exploration licence (Ryan 1991; Vail 1995; Biersack 1992). A simple version of the history of mining at Mount Kare is that Conzinc Rio Tinto of Australia (CRA) was granted the exploration licence in the area, but was prevented from proceeding by a massive indigenous gold rush in the late 1980s. At that time it is estimated that at least A$100 million worth of gold, mostly in the form of nuggets lying on or near the surface of the earth, was extracted, primarily by Huli and Paiela people (Jackson 1991; Ryan 1991; CRA Limited 1992). Huli men and women reminisce about the gold rush as a time in which 'money poured through our hands like water'; 'Air Niugini was our PMV (public bus)'; men would make grand entrances by helicopter onto the land of their future in-laws, pay brideprice in cash, and fly off again; and women could be 'like men' and 'go forward' in publicly contributing to relatives' brideprice payments, school fees, and pig feasts. While the squalor, disease, inter-clan violence, and prostitution that occurred during the gold rush has been emphasised in past literature about Mount Kare, what should also be emphasised is the generosity and intensification of social relations that the wealth made possible.

Many Huli people believed that the gold rush would never end, that the tension between giving and keeping was over for all time, and that a new era had begun in which both duty and desire could be satisfied. However, the surface gold did come to an end, abruptly curtailing the period of exuberant prodigality.

When CRA was finally able to proceed with the development of a mine, it came into conflict with local landowner groups, and it eventually gave up its lease to Mount Kare after a destructive raid on company property by local men (Vail 1995). Another mining company won the right to do hard-rock exploration there, but was challenged in the national court and only recently has been able to proceed (see Vail 1995 for a more detailed description of the legal history of the Mount Kare licence). So, after a wild gold rush which local people thought would never end, and during which everything seemed possible, there was a period of six years where nothing happened except complex litigation which almost no one understood. These legal obstacles have been exacerbated by the inability of local groups to achieve consensus on which 'landowner companies' and individuals may rightly represent them in their financial and legal negotiations. Moreover, at the time I did my research, the one landowner company which most Huli people agreed was legitimate – Karepugua Development Company (Ryan 1991; Vail 1995) – was on the verge of losing its alluvial mining lease because it owed the national government approximately A$200 000.

According to employees at the Chamber of Mining and Petroleum, expatriates claiming to be representatives of Mount Kare landowners had on two occasions tried to pay off this debt so as to appropriate control over the alluvial mining lease. Thus, unscrupulous entrepreneurs have attempted to exploit the confusion and conflict that has arisen over landowner representation, and the government has been left in the paradoxical position of threatening to cancel the alluvial mining lease unless the Karepugua Development Company (KDC) pays its debt, but being unable to accept any payment until it is absolutely sure that legitimate KDC representatives are the ones writing the cheque.

When Huli people discuss the recent history of Mount Kare, they often say that they are only living out what their ancestors predicted – and this is an important theme in contemporary discussions of myth. Traditional myths and names are said to be important now because they are actually 'prophecies' or 'parables' (Huli people use the English words) that people never understand until it is too late.[5] In this case, the name of the mountain, Kare, people say, is derived from the Huli word *'karere,'* which means to fight frantically over a desired item. It is what happens when someone puts out a bowl of food and tells people to help themselves instead of dividing up the portions for them. It is a situation in which the possession of some coveted item is suddenly undetermined and up for grabs, and it brings out the worst in people. So, people say, the ancestors who named Mount Kare knew that in the future it would be a site of greed, dissension, and turmoil. And clearly possession of, and authority over, the gold at Mount Kare has been problematic. Moreover, landowners in the Mount Kare area know that mineral development there has remained stagnant for about six years, largely because of the earlier conflicts amongst themselves and with CRA. As will be seen, this history is interpreted within a moral and mythological framework, but the issue of how to distribute blame – to the unyielding and monolithic *'kampani'* or amongst the fractious landowners – is an ongoing debate.

Huli discourse is saturated with references to 'the company.' This is perhaps not surprising since helicopters daily fly overhead between the Hides gas fields to the south of Huli territory and the Porgera goldmine to the north. Rumours abound in Tari that 'the company' is coming to build this road, or that office, or to distribute money for the use of various areas of land. Very few people seem to know exactly which company it is that is said to be coming, what they do, or why they have chosen to do what it is rumoured they will do. Nevertheless, when such rumours take hold, people amass the various documents and papers they have saved up over the years that might be important to 'the company,' and they eagerly wait near the airstrip for the designated helicopter or plane to arrive. Since 'the company' has come to represent and instantiate all that modernity or 'development' is and can provide, it is important to unpack indigenous notions of just what sort of entity the company is. This may best be done by comparing them with another group for whom 'the company' is salient – namely, bankers.

At a 'Mineral Exploration' conference hosted by the Papua New Guinea Chamber of Mining and Petroleum, all the mineral resource companies presented data about the sites where they were currently conducting exploration, and 'The Chamber' explained its capacities and its ideal role in the PNG mining industry.[6] The group was composed primarily of managers and geologists from the various companies, and the discourse was littered with talk of 'transfer structures,' 'porphyries,' 'mid-miocene arcs,' and 'intrusives.' The subject of landowners was confronted primarily through sarcastic asides, and for the most part was relegated to the category of 'logistical problems,' along with the lack of roads and the question of how best to manage mine sites if the new VAT tax was introduced the following year.

Interestingly, Rothschilds was well-represented at this conference, as a bank that supplies the loans necessary for 'project financing,' and they provided a cogent model for how 'the company' is conceptualised from the banking and investment end of things. To explore and develop a goldmine requires a vast amount of start-up capital, and it is always possible that the gold deposit ultimately revealed and developed will turn out to be much smaller than expected, so, to the bank, 'the company' is an entity which borrows a huge amount of money and engages in a high degree of risk to find, extract, and sell that commodity called gold, which goes to making jewellery, electronics, and to securing bank holdings. Moreover, while the price of gold has averaged approximately US$385/oz., the gold market fluctuates, and the spot price for gold is unpredictable. Therefore, 'the company' is also an entity that must be encouraged to 'hedge' – to sell a certain percentage of its 'proven and probable reserves' at a fixed middle price which provides security against a sudden drop in the price of gold but prevents the massive profits that are possible when the price of gold surges. Importantly, goldmining is 'country competitive,' and in a casual survey of Australian investment bankers about the degree of risk associated with various countries, Papua New Guinea ranked the highest. Therefore it is increasingly difficult for companies to attain the loans they need to develop mines in Papua New Guinea, and more and more exploration money is going to the 'emerging markets' in Asia and Latin America.[7] Bankers, the representative from Rothschilds concluded, 'are caught in a paradox of greed and fear,' constantly weighing the fact that the island of New Guinea is home to four of the ten largest goldmines in the world but is also considered the riskiest place for bankers to finance mines.

The one perspective that bankers and Huli landowners may share is that 'the company' becomes the focus for tensions between 'greed and fear.' From their experiences during the gold rush, many Huli men and women know what it felt like, however briefly, to be '*millionaire man na meri*,' and they want to feel that way again. However, it is 'the company' which has the technology to 'see' and retrieve the gold that is deep under the earth, and 'the company' elicits a myriad of conflicting emotions, from envious admiration to wariness and resentment, to a vulnerable hope for benevolence. The Huli attitude towards gold is a profound sense of entitlement mixed with an equally profound sense of dependency on 'the company' (Stürzenhofecker 1994). As in many Melanesian cultures, good fortune for the Huli is inextricably tied to the moral sphere. The fact that wealth is to be found on their territory is to them an indication of their moral and cultural superiority. Therefore, they deserve the gold that is to be found there. Moreover, they reason that they are 'the landowners,' and as such have pre-eminent claim over the land and any wealth to be found there (cf. Sagir, this vol.). As they see it, any company extracting wealth from their land should be grateful that they have given permission for this use in the first place and should therefore acquiesce to the reasonable demands for the construction of roads, health centres, schools, and so on.

On the other hand, they also talk about 'the company' as an alternative to the nation-state, alternately demanding or hoping against hope that 'the company' will step in where the state has failed in providing much needed social services and the promise of 'development.' In this kind of discourse 'the company' and the national government are spoken of as having equal status, but the company is said to have more money and to be better organised, and so it is desperately hoped that it will choose to take Huli people under its wing. In general the Huli see the company as a very wealthy entity that seems almost irrationally unwilling to part with its wealth and likely to be duplicitous, or at least secretive in a way that puts Huli people at a disadvantage. Huli men, for example, like to tell a story of how the Porgera mining company 'tricked' Enga landowners into signing away their land by treating a delegation of Enga men to a trip to Cairns. Huli men say that when the time comes and they are offered a free trip outside of PNG by 'the company,' they won't be misled by the temptations of a place like Cairns, but instead will demand to go to Singapore, which they say is where all their gold ultimately goes and which is now the 'true' source of power.[8]

In the case of Mount Kare there is another layer of discourse about the nature of 'the company' as a social or political entity, and this layer emerges from the specific history of CRA, both at Mount Kare and on Bougainville. CRA is a large and powerful mining company with sites in a number of areas in Papua New Guinea (Connell and Howitt 1991), most famously that on the island of Bougainville where there has been a long and bloody history involving secessionist rebels, government troops, and more recently a band of mercenaries secretly hired by the government to seize and reopen the mine site. Most Huli do not have detailed knowledge of the history of CRA on Bougainville, but what they do know has left people, especially men, confused about, and preoccupied with, the nature and degree of strength of 'the company' as a political entity. On the one hand, they know that the national government sent in troops against a group of landowners that came into conflict with CRA, and they feel woefully unprepared to deal with that contingency. On the other hand, they observe that CRA decided to evacuate their camp at Mount Kare after relatively little provocation from the Huli perspective. Some men gloat that 'we were able to run "the company" off,' while others feel that they overplayed their hand and would have done better to combine threats with enticements. They often assert that when 'the company' comes back to Mount Kare, 'this time we will get them to do what we want,' but whether that will be accomplished through threats, capitulation, or outwitting the company is unclear to them. Moreover, and more importantly, no clear 'we' has come into being. Indeed, the appearance of 'the company' as an innovative kind of corporate group that can expertly amass wealth has brought into glaring relief the problems Huli people have with mobilising collective action outside of warfare and, in the past, ritual. The myths and narratives discussed below demonstrate this current uncertainty about the possibility of collective agency.

Telling myths to 'the company'

One of the reasons the Huli feel entitled to the gold at Mount Kare is that traditionally the Mount Kare area was considered an important sacred site (Ballard 1994). It was inhabited and guarded by a shape-shifting python named Tai Yundiga or Tawa Tunduya, which is also the name of the sacred site (see also Biersack 1999). There are old men who can remember going with their fathers to Tai Yundiga to sacrifice pork there, and it is believed that the mythic python still inhabits and jealously guards the area. Tai Yundiga is described as a fearsome and demanding spirit. Men say, for example, that unlike other spirits, Tai Yundiga consumed the flesh of pork sacrifices, and not simply the smoke or smell as other spirits did. Thus, whole sides of pork had to be thrown into the sink hole where he lived. Moreover, when sacrifices were made, only bachelors who were well-versed in the necessary rituals were allowed to touch the pork; sexually active men were forbidden since Tai Yundiga could 'smell' the female substances on them. The presence of women was out of the question. Men often still say that it is inappropriate for women to be in the Mount Kare area, while women tend to emphasise that it is sexual activity that Tai Yundiga abhors, and not women themselves. Therefore, while men tend to blame disease and misfortune at Mount Kare on the presence of women, women argue that if people would simply refrain from sex while at Mount Kare then all would be well.[9] Tai Yundiga is also said to be the apical ancestor of many of the clans in the north Tari Basin, and most of the genealogical studies done by consultants for CRA document how various Huli and Paiela clans in the area are descended from Tai Yundiga (Henton 1988; Biersack 1999); for example, among the Huli, many clans in the parish areas of Puyero, Heli, Haya B, Kawi, and Doma C trace their ancestry back to Tai Yundiga, and in particular one of his daughters named Libime (see Map 2 and the genealogies in appendix). Similarly, conflicts amongst clans in the Mount Kare area about the right to compensation often take the form of asserting a 'closer' genealogical relationship to Tai Yundiga. For example, the Doma, a quite small clan, tried to assert greater rights than the Heli, a huge clan which includes both Huli and Paiela people, on the basis that they were descended from the first born son of Tai Yundiga's great grandson rather than his second born son (see appendix).

Collecting genealogies and myths among the Huli was a very delicate and sensitive issue in the context of mineral development. There were both suspicions and hopes that I might actually be a spy for 'the company,' and men in particular seemed to have a difficult time judging how much and what kinds of information to tell me. On the one hand I could potentially serve as a conduit, conveying informing to 'the company' that might increase chances of getting land compensation money. On the other hand, the company might use this information in combination with their own technology to locate and exploit other sites of mineral wealth. This approach-avoidance dynamic was further complicated by the fact that I was female, and as such should not have access to detailed genealogical information and certain kinds of mythical knowledge at all.

One of the myths I ended up collecting explains how the Tai Yundiga python came to Mount Kare from an area south of Tari. Since it is a narrative about the ancestor of Mount Kare area clans, it is actually part of a *dindi malu* – a genre distinct from other stories more properly thought of as myths or folktales. *Dindi malu* often include brief biographical narratives about the various ancestors, as well as the more genealogical information about who bore whom and where. When recording genealogies, men often asked me if I wanted the long or the short version of their *dindi malu*, the longer versions usually embroidered by stories about various ancestors, ranging from more fantastical for the ones furthest in the past (she has three heads and sits cross-legged on the mountain) to general description for more recent ancestors (he went to high school in Mount Hagen and now works at the post office). And it is my impression from the few quite detailed genealogies that I was able to collect that there is usually a long, quite secret narrative about the apical ancestor detailing his travels from the south to the north which put him in contact with cultural groups outside of Huli territory, and explicating his bodily transformation from or into various kind of animals, such as serpents or birds. *Dindi malu* are thus a complex blend of genealogy and myth, and in the context of mineral development and land compensation, it can be quite difficult to attain both components. Since it is thought to be the means of attaining land, the knowledge contained in *dindi malu* is considered quite powerful and is meant to be kept secret. Therefore, many men do not actually know all the complexities of their own genealogies, and many men do know, but will not tell, or will tell for a high price and with deliberate generational and mythic gaps. One man, for example, was willing to tell me the quite elaborate story of the travels and transformations of his clan's apical ancestor, but was not willing to tell me the names and places associated with his descendants. As he put it, 'this information is like a knife. It is my weapon, but it could turn back and cut my throat.' More commonly, men would be willing to give me a descending list of names, but not the mythic and biographical stories associated with them. And, importantly, *dindi malu* are meant to be told by men to men.

The myth in question, however, was told to me, a woman, by a woman, and the context in which I learned this myth illustrates many of the problems surrounding 'traditional' knowledge in the context of mineral development. The woman stated that her father taught her this *dindi malu* because he was getting old and her brothers had all out-migrated and/or died. However, she was too young and couldn't learn it as well as her brothers would have, and found listening to all those old names and places tedious. Later, when she realised how valuable *dindi malu* were, she wanted to go back to her father's home and learn more from him, but by this time she was married and her husband refused to let her visit her natal home. She added that when delegates from the mining company came to Mount Kare, she tried to approach the table to recite this *malu* so she could 'get her name on the list,' but she was shy, and in any case couldn't remember the names and places that define *dindi malu* as a genre and make them worthwhile to the mining companies. The context for this story points to a number of troubling issues: the predicament of old men who want to pass on *dindi malu* but whose sons are absent; the belief that women simply do not have the capacity to learn this kind of knowledge; the structural impediments for women to learning this sort of knowledge, such as access to male natal relatives after marriage; and the difficulty women have in proclaiming themselves to mining companies which have decided to rely on *dindi malu* as the only legitimate way to establish claims to land.

While I was transcribing the myth/*dindi malu* told me by this woman with my male field assistant, an older man overheard us. He was irate that I should have learned this story, and worse, that I learned it from a woman who had told it as if it were a folktale rather than a genealogy. Eager to be sure that I had the 'correct' story, I told him that he was absolutely right, the woman had told the story in a sloppy and inaccurate way, and that he should tell me the proper version of this story to be sure I had it right. After some deliberation he told me that the woman had gotten the place name wrong – that Tai Yundiga came from the Kutubu area not the Hides Gas area – and then he recited a 'short version' of the relevant genealogy which did not clarify or 'correct' the story I had been told. He concluded by sternly saying that I was not to ask women for that sort of information ever again. Women, as any Huli man or woman will tell you, have three essential purposes and capacities: to have children, raise pigs, and tend gardens. Moreover, as men add, they do not have the capacity to understand or articulate other sorts of traditional knowledge, and any stories they tell that have to do with the ancestral past will be garbled, incomplete, and of no value.

Serpents and shit: myths about mines

The myth itself says that one day Tai Yundiga's mother decided that it was time for her son to marry, and so she found a wife for him. He himself stayed in the bush hunting, and after months and months his young wife still hadn't seen him. She became angry and suspected that the old woman didn't have a son at all and had only married her to have someone to help with garden work. She had observed that the old woman left sweet potato roasting in the ashes each day before they went to the garden and that it was gone when they came home in the evening, and she wanted to find out if it was her husband who came to the house each day. But her mother-in-law insisted on accompanying her everywhere and expressly forbade her to be in the house alone. Of course, one day she told her mother-in-law she was going to fetch water and instead went back and hid in the house. As she peeped out the door, she saw a beautiful man striding out of the bush. His skin was glowing, his muscles were huge and he was adorned with leafy decorations found only in the bush. The girl guessed that this was her husband and was overjoyed that he was so manly and handsome. But as he got closer to the house, his body started changing into that of a python. As he entered the house he turned completely into a python and wrapped himself around the girl so tightly that they could not be separated.

By this time the mother-in-law realised that the girl had tricked her, and she ran back to the house to find her python son and the girl inextricably intertwined.[10] She was furious and yelled that this was not supposed to happen. Then she dumped them into a huge string bag and carried them all the way from south of Tari up to Mount Kare. Everywhere that she stopped to rest along the way – and more specifically, everywhere she sat (*'here winigoria,'* literally, where she put her buttocks) – special trees and plants grew; these are the spots where mining companies are now finding gold and oil. People add that the power pylons that go from Hides Gas south of Tari up to the Porgera mine in the north now mark the path the woman took. When the old woman reached Mount Kare she dumped the intertwined couple out of her string bag and they turned into two pools of water. Tai Yundiga is said to inhabit and, in a sense, to be the land at Mount Kare. People say, for example, that Tai Yundiga will rise up from where he has been lying, and thereby change the entire layout of the landscape, causing men in the bush to become lost. And, as Jeffrey Clark found during his research of the Mount Kare gold rush, the gold at Mount Kare is often said to be the skin and/or the shit of the Tai Yundiga python.[11]

What does it mean for the gold at Mount Kare to be thought of as the skin or shit of the powerful mythical ancestor? This belief encompasses a multitude of somewhat contradictory beliefs about both the power and the absurdity of gold; specifically, gold as shit is worthless, but gold as a product of the apical ancestor is powerful and, more important, it belongs to the Huli. As the outcome of personal consumption, faeces symbolise lack of social connectedness, and such disconnectedness, like faeces, is considered abhorrent and of no value. For example, when Huli children are given food by non-family members they are sternly told that if they did not participate in such give-and-take relationships they would have only their own shit to eat. To be human is to engage in exchange; to be non-human is to eat one's own shit (see Clark 1994 for a similar argument). Thus, gold is a substance associated with greed, asociality, and a condition which is repugnant, untenable, and absurd. Interestingly, mineral resources are symbolically associated with faeces in other areas of Papua New Guinea as well. The petroleum found in the Lake Kutubu area is in one Foi myth said to be the shit excreted by a white man after he ate a fish which had swallowed a coin he dropped into the lake (Weiner 1995a). In his analysis of this myth, Weiner points out the 'parthenogenic origin' of the oil: the white man produces this wealth item not through any sort of bodily or material exchange with others but through personal consumption, an act that has no social efficacy for the Foi. Weiner suggests that this myth is an implicit commentary on the 'isolated completeness that is the Western individual and his/her wealth objects' (1995a:164). In essence the Foi are saying that white 'wealth gets shitted away ultimately because it is never circulated and always privately consumed' (Weiner 1995a:167).[12]

And yet, in the Huli version of this shit-to-wealth alchemy, the gold is not the parthenogenic waste of the white man – it is something that has been digested and expelled by an ancient, powerful, mythic ancestor from whom all the Mount Kare landowning clans are descended and to whom sacrifices were regularly made until recent times. What kind of meanings does this particular sort of faeces have for the Huli? On one level this belief is simply saying that while gold may be an asocial substance that indexes a kind of subjectivity and economy that is totally 'other,' it nevertheless belongs in the deepest possible way to Huli landowners. However, it is also important to keep in mind that in the context of myths and genealogies about apical ancestors, shit often reveals itself to be something else. For example, in the founding myth of one Mount Kare based clan, it is revealed later in the genealogy that what had earlier appeared to be a pile of excreta was actually the heart of the clan's founder. Similarly, in the above myth, it is said that where the woman 'put her buttocks' new species of trees and plants grew. When I asked for an explication of this imagery I was told that the woman was simply stopping to catch her breath, but given the Huli propensity for euphemism (Goldman 1983), this image could quite easily refer to her faeces. In many Huli myths mythical characters metamorphose into other substances such as mud or shit to escape the human protagonist. Thus, shit is not always transparently symbolic of that which is worthless and abominable. It often disguises that which is most authentic, generative, ancient, and valuable.

When one contextualises the Mount Kare narrative in other Huli myths and beliefs, a more complex meaning is exposed. While the serpent in the Mount Kare myth is quite sexual, it is more common for the image of the serpent to serve as a symbol of purity, continuity, asexuality, and renewal. For example, the central feature in Huli sacred geography is a subterranean python stretching through Huli territory from south to north (Goldman 1983; Frankel 1986). This python is intertwined with a long strip of cane which holds it in place, and the survival of the world is said to depend on the python's body being kept in a straight line. Similarly, the most well-known Huli myth explains why snakes need only shed their skins to be renewed, while the lot of humans is death and all the conflicts and calamities that lead to death (Goldman 1983). This myth asserts that long ago the first man approached the house of his wife to give some bespelled water to their first child. He repeatedly called out to her, 'Mother of Life' ('*Habe Ainya*'), but the woman refused to answer. Only when he called out 'Mother of Death' ('*Homabe Ainya*') did she respond, and so, perhaps in despair or perhaps in disgust, he threw the water into the bush where a serpent consumed it and became immortal. The human child drank breast milk and was fated to mortal embodiment. This latter myth implies that in fact humans were meant to be immortal – they were meant to consume the water that would have endowed them with a better sort of bodily life. However, the contingent nature of human sociality, in which conversations and even whole relationships can go awry in the space of a moment, changed the course of human history. Due to an instance of female sullenness and male impatience (most Huli, both men and women, attribute blame to the former not the latter), the magical fluid ended up in the belly of a serpent, and now the serpent can repurify and revivify itself by merely shedding its skin.

Humans, on the other hand, are vulnerable to disease and aging, and moreover must risk the dangers and impurities of sexual reproduction to 'replace' themselves (Biersack 1995b, 1999). Because they are doomed to mortality, they can only achieve a sort of immortality through reproduction (cf. Wagner 1967). Sexual reproduction requires the exchange of women, which entails the ongoing exchange of gifts, and thus the entire nature of the Huli social system is determined by the fact of human mortal embodiment and the necessity of sexual reproduction.[13] For Huli men, one of the consequences of sexual reproduction is exposure to female sexual fluids, which can have a host of detrimental effects. These include the dulling of the skin and the possibility that one's intestines will twist into knots and that one's shit will eventually explode inside the body causing death. Beliefs about menstrual pollution therefore recapitulate the Huli serpent myth that explains the origin of human social life, and similarly oppose the bodies of men to the bodies of serpents. In the myth, the male child consumed a female substance (breast milk) and became mortal; in day-to-day life, men who consume, or are in any way exposed to, other female substances (sexual or menstrual fluids) risk death. Spells to prevent or cure this damage include allusions to serpents which can shed their dirty skin and whose intestines are smooth and straight (Goldman 1983). Serpents are symbolically the opposite of humans: they are immortal and thus do not have to replace themselves as humans do by engaging in sexual reproduction and all that sexual reproduction entails.[14]

The belief that gold is the waste of the mythical serpent engages this myth on multiple levels. Like the lost water of immortality, the gold at Mount Kare is perceived as a transforming substance meant for the Huli. And like that water, the gold is being consumed by others who were not meant to have it. Indeed, this myth is now sometimes called a 'parable,' or prediction, and history is thought to be repeating itself. 'The company,' it is said, is calling out 'Mother of Life' to the Mount Kare landowners, but they are acting like oppositional, refractory, and sullen women who refuse to answer when called.[15] Thus the gold is going elsewhere, and, like the mythic water, seems to take on value for the Huli only when it is consumed by others who serendipitously attain what the Huli lose through their wilful, fractious nature. Notice that in reformulating this myth as a 'parable,' Huli men are envisioned as feminised and with the worst of stereotypical Huli female characteristics: oppositional for no apparent reason, selfish, and myopic about goals that transcend the individual.[16] The proper behaviour in this imagined scenario would be that of a docile and amenable woman who responds to whatever name 'the company' wants to call her by.[17]

Fantasies of agency

Men and women differently draw on these mythic themes of serpents and faeces to construct fantasies about people who exercise untraditional forms of agency and thereby come to own and control the gold at Mount Kare. What I am calling fantasies are essentially stories I heard from Huli men and women when I asked about Mount Kare or other sites in the Tari Basin where gold or oil was rumoured to have been found. While these narratives could perhaps be called contemporary myths, this may not be the best way to conceptualise them. Huli people do not think of them as traditional myths – they do not account for the origin of anything, do not take place in some deep past, and they do not fit into the current reformulation of many myths as 'parables.' Moreover, they are not part of any stock of well-known stories at this point, and precisely what they are as a genre depends on one's perspective. What I call the female fantasy is dismissed by Huli men as women's nonsense, and perhaps would be called fantasy if the Huli had such a 'genre.' Women, on the other hand, claim that such stories tell of actual events that have happened to women they know, or at least know of. What I am calling the men's fantasy is at the current time hoarded by a few men as secret knowledge, but since it is believed to have happened to a particular man it may eventually become part of a *dindi malu*. Within Huli epistemology and classification of narratives, the men's fantasy would – even by women – be considered more valuable and legitimate. However, as will be seen, the stories are very similar in both structure and content. It is therefore a deliberate strategy on my part to equate the two stories by giving them the same label.

It is also a deliberate strategy to combine the notions of 'fantasy' and 'agency,' a combination that seems counter-intuitive. Fantasy connotes interiority and unconscious or irrational projection, while agency implies intention and deliberate action in the world (Knauft 1995). However, in the stories I am about to discuss, desire links directly to pragmatic action and outcome: individuals want wealth and they forge specific kinds of relationships and engage in specific activities to obtain it. The seemingly fantastical and the practical play equal parts in these narratives, and they are considered, by the people who tell them, to be legitimate evidence of authority over sites of gold and oil. In his analysis of similar narratives, Lattas, following Taussig (1980, 1987), interprets Kaliai cargo cult stories as 'the imaginary structure of the colonial encounter, that is the role fictions played in mediating the colonial encounter' (Lattas 1992:28). The Huli narratives discussed below are similarly responses to a specific type of (post- or neo-) colonial encounter, namely that with 'the company' and the world of multinational mining, and they go towards creating a separate fictional realm where this encounter plays itself out differently. However, these narratives are a specific kind of fiction: they often begin as an individual's dream which is then publicly shared, interpreted and elaborated, and they are permeated by overlapping and entangled notions of desire: sexual desires merge with consumption desires, which become mapped onto desires for power and collective solidarity.[18] Finally, in line with a more psychoanalytic notion of fantasy, these narratives use symbols – particularly that of the python – to restructure the company/landowner relationship, splicing economic and cultural antagonisms into a seamless, unproblematic story about the Huli production and possession of gold.

There are a number of variations of the female story of fantastic agency, but the essential narrative goes like this:

'The company' found gold on my clan's land and this is why. There was a young woman from my clan who was walking along the road alone. A handsome man who looked like he had just stepped out of the bush approached her and 'pulled' her. (This word 'pull' – the *Tok Pisin* word *pulim* – can mean anything from briefly pulling on someone's arm to abduction, but it usually has sexual connotations.) The man turned into a python and wrapped himself around her. Then he took her down a hole to his underground city where there were permanent houses that had electricity and chairs and verandas. Everyone travelled around town in Toyota Landcruisers. He told her, 'I am Tai Yundiga. The people at Mount Kare were so greedy for my gold that they not only dug up my shit, they also crawled up my anus and were digging out my intestines. So I ran away underground and have come to your clan's territory. I want to marry you, and in exchange I will show you where my gold is. You cannot tell anyone because I am looking for a place where I can be at peace.' Then he let the girl go home. She is his wife and sometimes he takes her away with him to his city, but usually she just lives with her parents. That's why they have discovered gold on my clan's land.

The story I heard from men is less well known – it is probably only known to men who can claim to be Mount Kare landowners – and to some extent is treated as secret knowledge, similar to the ancestral genealogies to do with Mount Kare. In this story a man is alone and lost in the bush near Mount Kare. He walks and walks, and notices that all around him are the bones and relics of a civilisation prior and ancestral to that of the Huli. He finally comes upon a paved road, and eventually a Toyota Landcruiser comes up behind him. He sees that the driver is his cousin who died years ago. He hops in and is taken to a fenced-in block of permanent houses, all of which have electricity. There are no women at the camp, and the man realises that all the men are Huli and Paiela relatives of his who have died. He goes to the mess hall and eats with his cousin who says the men there work in shifts and are paid fortnightly. The man begs to be shown what sort of work it is that they do there and his cousin is reluctant to tell him, but finally agrees. The next day his cousin takes him to a deep open pit. Down in the pit the men are holding a huge rubber pipe. Something is flowing through the pipe, and the pressure forces the pipe to twist this way and that. The men's work is to keep the pipe straight and to hold it off the ground until the next shift of men comes to take their place. If the pipe slips from their grasp and falls to the ground or twists out of control then the world as we know it will come to an end. The cousin then takes him back to the bush and tells him that as long as he never reveals the secret things he has been told, he will soon gain wealth from Mount Kare. This man was the first to see the company helicopter flying overhead and the first to find gold at Mount Kare. And he never did reveal the secrets he was told – all that I have told you here is that which he was given permission to discuss.

These two fantasies are in many ways the same story. Both start with a lone person who is approached by a mythic figure on the road. Both use the image of the subterranean serpent intertwined with, and held in place by, something else – in the traditional Huli cosmology it is a length of cane; in the women's fantasy it is a woman's body that is entwined with the serpent; and in the men's fantasy it is men's hands holding the rubber pipe or snake in place. (When I pointed out to my field assistant that this rubber pipe seemed a lot like the Huli subterranean python, he told me that I had probably guessed one of the secret things the man was told never to reveal.) Both gain entry to a Western-style city with trucks and permanent houses, but it is implied that this urban space is part of an ancient mythic underground landscape and therefore prior to and more genuine than any actual Western city.

The three themes that these stories have in common – the road, the underground world, and a relationship to the python – are important for understanding these fantasies. In both traditional myth and in the current Tok Pisin vernacular, the road or path is symbolic of choice and personal transformation. Traditional Huli narratives, both folk tales and the more prestigious and 'true' *dindi malu*, often begin with the protagonist setting off on a journey or becoming lost on the road where s/he meets figures who pose riddles, demand that choices be made, or give advice which may or may not be followed. While on the road things are never what they seem, and journeys always transform these protagonists. Moreover, the *Tok Pisin* word for road or path also denotes alternative life choices and ways of being. For example, Huli men these days are said to have to choose between '*rot bilong tumbuna*' (traditional ways) and '*nupela rot*' (modern ways). The symbol of the road in both of these fantasies therefore implies agency and transformation. Similarly, Lattas has pointed out for Kaliai cargo cult myths that underground mythic spaces often denote hidden truths and realities not yet discovered or colonised by whites: 'The visible everyday world of meaning becomes a surface which is explained through the animating powers of a less visible world' (1993:68). Thus the image of the underground draws on pan-Melanesian metaphorical oppositions between the skin or surface and the bone or root in which the underground realm is associated with that which is causal and true, while the surface can be deceptive and misleading. And, finally, both the woman and the man gain exclusive and legitimate access to gold by being able to establish fundamental and necessary ties to the ancestral python. Importantly, these are ties that are necessary to the python. In both fantasies it is the python which is vulnerable and requires Huli assistance: the python needs safe haven; it needs to be held above the earth. Thus both fantasies indicate entry into a hidden but 'true' world where it is the fate of the mythic ancestor that determines what will happen in the above world, and the actions of Huli people determine this fate.

But there are also important gendered differences between these two fantasies, and these differences turn on the forms of agency used to establish a tie to the ancestral python. The keys to the male fantasy are labour and the male corporate group: the men hold the python-like pipe with their hands, they work in an open pit in shifts, and they receive a fortnightly salary. In representing the mining camp as a group of male ancestors, this fantasy confronts anxieties about male collective potency and the role of the clan in the control over new forms of wealth and power. Agnation and male solidarity, once so necessary for warfare and ritual, are here re-imagined as a team of wage labourers. However, this is wage labour with a much higher cosmological purpose – by holding the pipe in place these men are saving the world from destruction.

In this exclusively male realm – just as the ritual site for Tai Yundiga was meant to be – Huli male ancestors have already imagined the landscape of wage labour. They, in fact, own and authorise wage labour at Mount Kare, but they have invested it with mythic significance. The men gain special and exclusive access to the wealth of Mount Kare, but it is because they are performing important cosmological work; thus, the clan as a potentially obsolete organisational form is refigured as a new and efficacious one that successfully maintains control over the wealth at Mount Kare. The narrative draws on both phallic and anal imagery to show that the flow of wealth and the continuation of the world are dependent on male cohesiveness and clanship as represented by the communal

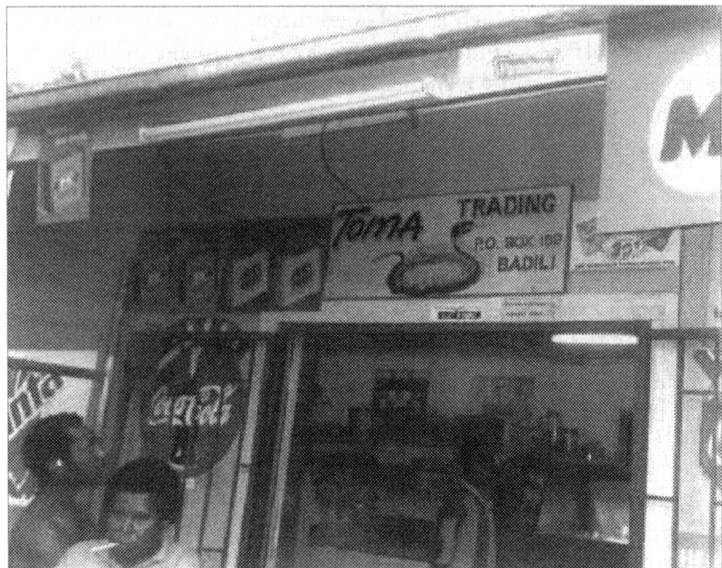

Plate 1: Huli-owned trade store in Port Moresby, with logo containing clan name and image of the Mount Kare python.

phallus which they all hold in their arms, and on male purity as represented by the smoothly flowing waste of the serpent. Moreover, the representation of men in this fantasy explicitly refutes the *Habe Ainya/Homabe Ainya* (Mother of Life/Mother of Death) myth-turned-'parable' in which Huli men behave like recalcitrant and oppositional women while 'the company' acts as the true embodiment of masculine solidarity and purpose. Here clan cosmological potency is equated with sexual potency which becomes clan economic potency, and indeed, the image of the Mount Kare python has become an important symbol for landowning clans. This is evident even in Port Moresby where the trade stores owned by Huli men often use a logo combining the clan name with the image of the python (see Plate 1).

Interestingly, while in this fantasy the clan is re-imagined as a team of labourers, wage labour is in fact not yet a significant aspect of life at Mount Kare. However, Huli men have experienced wage labour at mines such as Ok Tedi, Porgera, and even Bougainville. It is typical for Huli men who have held a job to say, '*moni, em gaden bilong ol waitskin, na graun, em moni bilong mipela*' (money is the garden of white people and land is our money.) In other words, there is an explicit attempt to make sense of wage labour and money as an alternative form of production and of life, and there is an explicit sense that one must make a choice. As Huli men see it, those who depend on the land do not have easy access to cash, but they are 'free' (Huli use the English word) – free to sleep, eat, and work when they want – a dimension of 'traditional' life that is highly valued. Those who have jobs have money, but are not 'free' – they must stick to a routine and obey others' orders. And while land and money are both forms or sources of wealth, they are not easily convertible. Land sustains indigenous people, and money sustains white people, and yet there is no clear connection between land and money. However, in this narrative, wage labour, in which one is alienated from one's labour and the means of production, is here fantasised and redeemed as a kind of necessary intimacy with the earth and with the ancestor who embodies and safeguards it. Gold becomes that heretofore hidden connection between land and cash: gold is born out of the land and is dependent on a cosmological tie to the land, and yet it can be traded in for vast sums of money.

However, this fantasy also expresses tensions and anxieties about the mining process. Since the narrative fuses the ancestral python with the company pipeline, it poses a fundamental question: if the python and the pipeline are the same, which one is it really? Is the company pipeline really our ancestor, or is our ancestor really the company pipeline? Is the discipline these men display the sort of ritual discipline and purity required by Tai Yundiga, or is it the discipline mandated by company employers? Is our male collectivity producing wealth that is meant to be ours, or is this company going to shit all over us? It is implied that the necessity to support the ancestral python is a result of the fact that men are removing the earth which is meant to support it. Thus, as men above the earth dig away at Mount Kare in their greedy and selfish quest for gold, the ancestors beneath the earth must work ever more strenuously to hold the python in place, and the control they have over the ancestral figure/pipeline which sustains the world becomes ever more tenuous. Thus the value of mining and the fetishisation of gold is thrown into question (Taussig 1980).

Not surprisingly, the female fantasy does not engage notions of collectivity or the meaning of 'the company' in the way that the male fantasy does. In the past women did not represent the clan as a corporate political unit, and currently women do not act as landowner representatives in interactions with the company, nor do they go off for wage labour at mining and oil drilling sites. Female agency has always expressed itself more atomistically among the Huli. Moreover, their experience of the Mount Kare gold rush was quite different from that of men's. Many women only experienced the gold rush through the stories brought back by their husbands and brothers. Men came home with marvellous tales, some of which had to be lies, women told me, but they didn't want to be made fools of by men, and so they kept ambiguous smiles on their lips which could either be interpreted as knowing derision or admiration. Who could believe, for example, that nobody walked on foot in the capital city, or that books and magazines were actually made out of trees? One kind of story, however, turned out to be true. There were in fact *pasindia meri* (literally, passenger women or women on the move, prostitutes) at Mount Kare who were having sex with men and getting money for doing so. And many *wali ore* (literally, real women; good women) who made the trip found themselves transformed into *pasindia meri* when they entered the space of Mount Kare where suddenly anything seemed possible. It was also true that women returning from Mount Kare often came home with vast sums of money, from prostitution or from panning for gold, and were able to contribute publicly to relatives' brideprice, compensation, and school fee payments. Women who used their wealth in this way were praised for being 'like men' and for 'going forward' in promoting family and clan interests.

The female fantasy engages both of these elements. For women the access to gold and all it represents is clearly through sexuality and marriage with the creator and guardian of the wealth. The mythical python chooses individual women and offers marriage so as to gain rights to an area of land where he can find some tranquillity. This narrative is striking on three counts. First, the narrative structure of the story itself very closely resembles women's life history stories of how they first met their husbands or first had sex. The typical story of a woman's first sexual encounter tells of being spied on by a young man while working in the garden and then being accosted by him as she walks back home alone. These narratives are always erotically and emotionally charged, and many women have a hard time deciding whether sex in these cases is consensual or not. Many women claim that force was used and that they were threatened with knives. Nevertheless, all women would agree that to walk alone on the road is to 'ask for it.' While this notion of 'asking for it' simply by walking alone may strike the reader as an ideology that controls female sexuality – and in many ways it does – nevertheless, many women perceive it as a way of exercising a kind of passive sexual agency: women can choose to make themselves available without having to say so in so many words. And, as long as one is accosted by one's intended suitor (which is not always the case), there is a certain sexual charge to be had from being 'pulled.' The possibility of being accosted or 'pulled' by the Mount Kare python is quite real to women. 'Rape by snake' is listed in the Tari Hospital medical records as a reason for admission along with such items as spear wounds, NVD (normal vaginal delivery), and severe anaemia (see Plate 2, overleaf). Therefore, while the two fantasies may look structurally similar, it is important to bear in mind that a Huli woman walking alone on the road is not at all the same as a Huli man walking alone on the road. The fact that the female protagonist is walking by herself implies that she is inviting some sort of sexual encounter, or at least that she is enticingly and provocatively alone.

Moreover, this fantasy is striking for what it doesn't say as well as for what it does. Importantly, being chosen by the python and thereby gaining access to all his wealth does not require reproduction or physical labour. None of the three capacities intrinsic to and expected from Huli women – raising children, pig tending, and garden work – are demanded. Rather, the woman is simply desired by the python, and that is sufficient for a proposal of marriage. And, also important, the python comes and lives on her land and allows her to stay and share her wealth with her natal family – a marital situation many Huli women would describe as ideal. Finally, it is striking that in this fantasy women are valorising that which men would consider completely taboo given traditional beliefs about Tai Yundiga. Men like to emphasise that women should not be in the Mount Kare area and that their presence offends Tai Yundiga, causing disease and accidents. Here, women seem to subvert such beliefs by reasoning that if they cannot go to Tai Yundiga to get his wealth, well then he will just have to come to them. In this fantasy women are playing with a new form of agency that is becoming increasingly salient in the current socioeconomic context, specifically sexual agency. Melanesian ethnographers have pointed out the changing sexual scene in Papua New Guinea (Jenkins 1994; Zimmer-Tamakoshi 1993; Hammar

PAPUA NEW GUINEA
DEPARTMENT OF PUBLIC HEALTH
HOSPITAL OR HEALTH CENTRES

ADMISSIONS

PERIOD FROM

Name	Place of Residence	Provisional Diagnosis
LEGYO H	TARI	CONFINEMENT
PENDO A	"	"
JOJINGA T	"	HN SEPSIS
MARY M	"	T/ABORTION
HOYAME I	"	MOD PNA
LEGA LUGU	—	KNIFE WOUND
SYBILI J	TARI	N.V.D
AGNES P	"	"
FABAL L	"	"
IRENE M	"	MOD PNA
PUTI L	—	RAPE BY SNAKE
MARY K	TARI	BREECH DELIVERY
EKI ZOMBE	"	SEV. ANAEMIA
KEPEN H	"	SPEAR WOUND
WANGO K	—	MALARIA/PNA
DITO J	TARI	ABORTION
TAI MEGIA	"	CONFINEMENT
ALBERT M	—	S. PNA/TB

Plate 2: Admission page from Tari Hospital.

1992; Nihill 1994; Knauft 1997), and it is becoming clear that in the postcolonial context people are constructing modern identities through experimenting with new kinds of sex.

While the female fantasy may not be a response to, or reinterpretation of, interactions with 'the company,' it does engage racial tensions of the postcolonial encounter. Indeed for women, it seems that the python has shed his skin again, and now it is white. When I asked women what the Tai Yundiga python was like, some told me that in dreams he had appeared to them with white skin and fine, straight hair. It is important to point out that the idealised person for a Huli woman to establish a sexual liaison with, or better yet a marriage with, is a white man. White men have the reputation of being wealthy, kind, and relatively willing to treat women as companions. Moreover, the underground urban space imagined by women in the female fantasy is distinctly different from the male fantasy. Where men imagine company dormitories and canteens, women imagine verandas and chairs and a kind of 'white' sociality in which men and women sit on their porches and drink tea. If for men the symbol of the python represents a tension between clan collectivity and the kind of unity and power embodied in 'the company', women draw on the more sexual aspects of the python and imagine themselves 'taken' by, and into, a modernity where wealth and whiteness is attained simply by being desirable.

Conclusion

Gabriele Stürzenhofecker has pointed out for the Duna that in the past ritual sacrifices of pork were used by people to tap into and manipulate power deep within the earth. However, as she puts it, there has been a 'replacement or deflection of ritual by technology' (1994:39); now it is the mining companies, and the tools and knowledge they have, that tap into and manipulate the earth to gain wealth. Fantasies of agency are about suturing modern desires and powers to those powers and desires which are most historical. The past cosmological influence over the earth and its fertility is simply transposed into the desire for, and control over, gold. Instead of a ritual tie to the earth and its mythic guardians, people imagine other fundamental ties that would once again give them direct access to the earth and its wealth. I say 'once again,' but these fantasies are potent in part because the figures in these stories are imbued with a certain primacy by being anchored in traditional mythic images. The Huli see the gold going elsewhere and transforming others into the *millionaire man na meri* that the Huli were meant to be, and they imagine new ways of acting in the world so as to regain control over the wealth they believe is rightfully theirs. Men dream of a collective agency that would rival that of 'the company,' while women dream of gold as the most ancient and valuable brideprice a woman could bring to her family. At the same time, these fantasies engage tensions associated with non-traditional types of agency. Men valorise wage labour as cosmologically meaningful work; women use a dangerous kind of sexual agency to seduce the apical ancestor. By reappropriating the future and fusing it to the mythical past, people smooth over the rupture between past and present and the trauma of alienated land, company discourses and agendas they do not understand, and the dissipation of wealth to other inaccessible sites of power.

Acknowledgments

The research for this chapter was funded by grants from the National Science Foundation (9412381), Wenner Gren (5848), Fulbright Hays (PO22A40009), the Association for Women in Science, and the National Women's Studies Association. Comments are gratefully acknowledged from Bruce Knauft, Don Donham, Dan Smith, Gayatri Reddy, Andrew Cousins, and everyone who participated in the 1997 Myth to Minerals conference.

Appendix

1. Descent of some north Tari Basin clans from Tai Yundiga

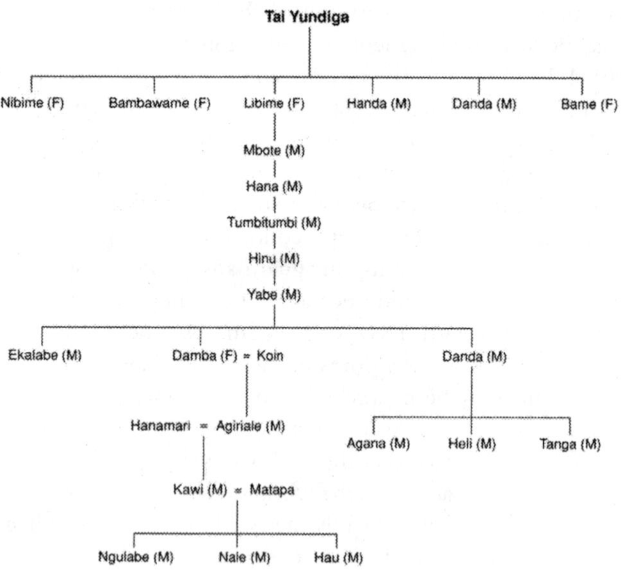

Figure 1: Descent of some north Tari Basin clans from Tai Yundiga.

Note: This genealogy and the following do not in any way attempt to be comprehensive or to account for the descent of all clans who are now claiming to have rights to Mount Kare. These genealogies only cover in a very schematic way a few of the Huli clans who live in the north Tari Basin and are said to be descended from Tai Yundiga. This information is based on just a few genealogies, and it is intended only as an example of how local clans trace their descent from Tai Yundiga. This particular genealogy was given by a Hau man, and therefore reflects a Hau bias. Hau, Ngulabe, Nale, Agana, Heli, and Tanga are all clans in the Paiyaka area

Map 2: Parish areas in the north Tari Basin.

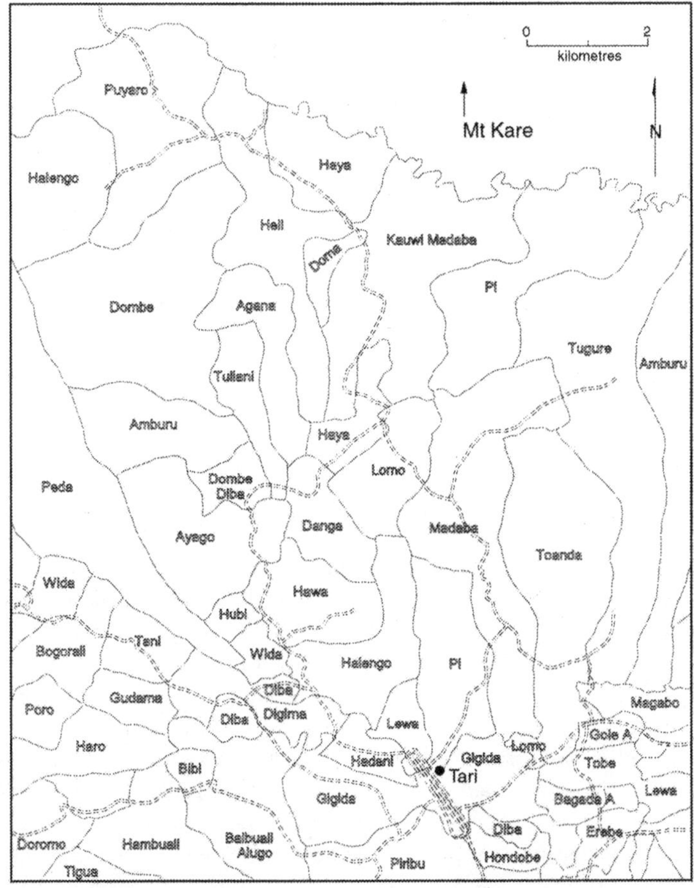

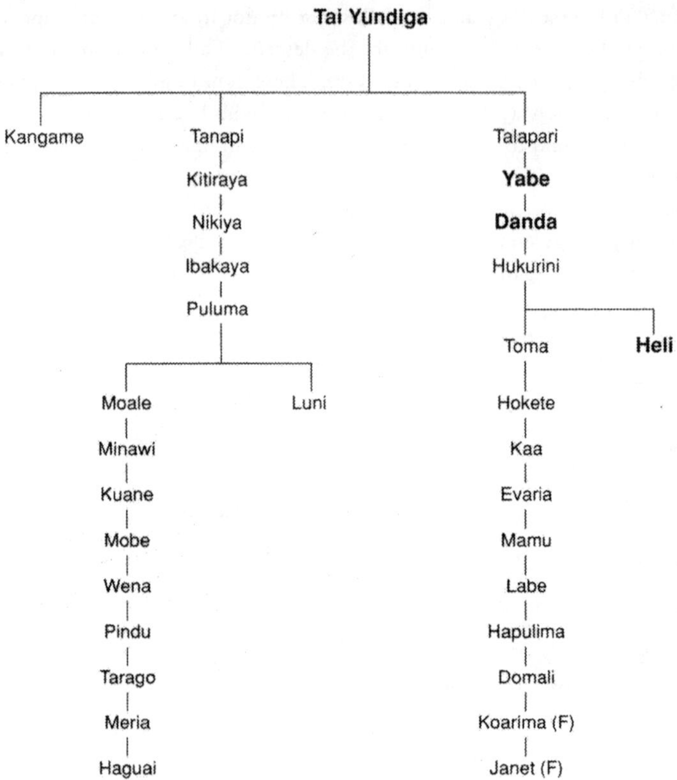

Figure 2: Descent of Puyero and Toma clans.

of the north Tari Basin (Hau, Ngulabe, and Nale are subsumed under the umbrella name Kawi – see Map 2, page 61). All are descended from Tai Yundiga. The Heli are generally considered the rightful owners of the Mount Kare area because they traditionally carried out rituals there.

The links above do not necessarily represent parent-child relationships since there are a number of generations missing from the diagram. For example, see the next set of genealogies for the Puyero and Doma (or Toma), clans also descended from Tai Yundiga and also acknowledged to have rights to compensation for use of Mount Kare land. The Huli word for 'bore' (*honowini*) does not in genealogies refer to an immediate reproductive relationship, but rather to being the forbear of somebody. Thus Tai Yundiga can in the discourse of *dindi malu* be said equally to have 'born' Libime or Yabe since he is the ancestor of both of them. This means that when men are giving cursory descriptions of their genealogies they will often collapse or bypass generations so that quite distant relationships come to look like father-son relationships, and father-son relationships can appear to be brother-brother relationships – for example, Tai Yundiga bore Danda and Heli.

2. Descent of Puyero and Toma clans

See Figure 2, page 62.

Note: Individuals also represented in the first genealogy are in boldface. It is not clear to me why Libime and her siblings, most often cited as the immediate offspring of Tai Yundiga, were not articulated in this genealogy, or how Tanapi, Kangame and Talapari and meant to be related to this other sibling set.

While only the lineages of two individuals are represented here, all Puyero clans are said to be descended from Puluma, and all Toma clans from Toma. This genealogy is greatly simplified since wives and siblings are not represented. Other branches of Puyero and Doma clans are descended from other offspring of Puluma and Toma. Haguai Meria and Janet Koarima are two individuals who have been very active in Mount Kare mining negotiations. Janet Koarima is currently president of the Tari District Women's Association.

Notes

1. I have coined this phrase on the model of one used in chapter 3 of Henrietta Moore's book, *A Passion for Difference*, namely 'fantasies of identity' (p.63). In this chapter 'fantasies of agency' are more precise in that people are not imagining whole new identities but rather new ways of acting in the world.
2. I say 'satanic' because many 'traditional' or non-Christian beliefs and practices are now often labelled '*samting bilong Satan*' (Satanic).
3. Clearly I was chosen because I speak English, was living on the land of the clans who were forming this landowner group, and am white, which connotes power, wealth, and connections to the outside world. I was not given any position of authority, and the group was short-lived.
4. Nellie James said it was possible to imagine a situation in which landowners could attempt to reject mineral development on their land; for

example, if they could demonstrate significant 'biodiversity' on their land that could not be found elsewhere. However, they would have to argue this in court, and, in any case, such an incident had never occurred since seemingly everywhere people in PNG are eager for the 'development' and cash compensation that comes with mining companies.

5. People also sometimes say that myths are real events that happened in the deep past as well as 'prophecies' or 'parables' for the current time. The fact that myths are now seen as cryptic prophecies reflects: a desire among the Huli for their narratives and metaphors to be as powerful, revealing, and unifying as they perceive Christian texts to be; and a sense that they have lost cultural knowledge, and do not understand some of the metaphors and turns of phrase that older people use.

6. The Chamber of Mining and Petroleum, along with the Geological Survey, ideally plays a number of important roles in the Papua New Guinea mining industry, from providing high-tech geological mapping services to supplying 'wardens' for meetings between landowner and company representatives. However, the national government annually cuts the budget for these two institutions, preventing them from offering many services at all. Moreover, almost all Papua New Guinea graduates from the University of Papua New Guinea geology department are quickly employed by the various mining companies in the country so that 'The Chamber' is hard put to find capable employees.

7. Indeed, while anthropologists and environmentalists may worry that too much mining is being done too quickly, mining company and government representatives are concerned that not enough mineral exploration is taking place and that there will soon come a time when existing mines are depleted and no new mines have been developed, leaving the country without a critical source of income.

8. This belief may have emerged because when Huli men went to banks to exchange their gold they were often told to come back later when the banks had received the spot price for gold from Singapore.

9. Jeffrey Clark reported that men he interviewed at Mount Kare during the gold rush suggested that gold itself was like menstrual blood and was polluting (1993). Like Biersack (1999), I never heard anyone draw an analogy between menstrual blood and gold. However, as this chapter discusses, gold is sometimes said to be the faeces of Tai Yundiga, and, in some contexts men also associated menstrual blood with faeces. Indeed, this is how young boys are euphemistically taught to keep their distance from women and women's houses: they are told by older men that women never adequately learn hygienic practices for excretion and inadvertently

get faeces under their finger nails, which then spreads to their houses. Only upon marriage, when it is appropriate for them to learn about female sexuality and reproduction, do they learn that this initial information was wrong. Perhaps the common denominator, then, is faeces, not menstrual blood.

10. He wrapped himself around her so that they were intertwined head to head and thigh to thigh. He wrapped himself so that she was underneath him and he was on top and his tongue was flicking through her black hair: '*Mandamanda* (head to head) *piya ke pagopago* (thigh to thigh) *wai puya piagome hibi hupa piya* (wrapped). *Wandari piago mo handa taga* (underneath) *hayiya; puya piago taliga* (on top) *hayiya. Popuyalu popuyalu* (wrapped), *manda mindini* (her black hair) *puya piago ibu hege tomatoma* (tongue flicking) *pialu pere laya.*'

11. This myth resonates with myths from a number of other culture areas in PNG. Among the Telefolmin, the travels of the female mythic figure, Afek, also had propagative effects on the landscape (Brumbaugh 1990), and Magalim, another key Telefol mythic figure, often takes the form of a python, is associated with pools of water, and is now said to haunt the company canteen at the Ok Tedi mine (Jorgensen 1980, 1993). Malila, the key mythic figure for the Maenge of East New Britain, also appears to mankind in the guise of a python, and one myth tells that he was carried in a basket by his mother, who shaped the landscape as she walked along. Moreover, ritual garden magic among the Maenge is said to use 'snake shit' (Panoff 1970). The Kaliai of West New Britain tell a myth very similar to that of Tai Yundiga in which a woman becomes suspicious of her snake-man husband because she never sees him, and brings about her own downfall by spying on him (Lattas 1992:39). It is possible that the Huli learned these myths from other cultural groups and incorporated them into their own. Among Papua New Guineans, the Huli have a reputation for being peripatetic; moreover, when pacifying the Highlands, the Australian colonial government deliberately tried to form the Southern Highlands as a labour pool for coffee and coconut plantations in other provinces, making it more likely that they would be exposed to the beliefs of other peoples (Strathern 1982; Harris 1972; Stewart 1992).

12. Moreover, the white man returns and builds a pipeline, an image that in the present context can be seen as intestinal, which carries the petroleum back to his home, thereby consuming his own shit. Thus, in this myth 'white' wealth is not only 'parthenogenic,' but also partheno-phagic – precisely what Huli children are told is their dreaded fate if they fail to

realise how deeply they are embedded in, and indebted to, the web of exchanges that has created them.

13. To attribute the nature of the entire Huli social system to this one myth is not an exaggeration. People say, for example, that if sexual reproduction were unnecessary then men would not fight or take each other to court, since many conflicts are ultimately caused by women.

14. Biersack, in her recent interpretation of Mount Kare serpent symbolism, argues that Tai Yundiga, as the apical and totemic ancestor to whom local clans made sacrifices, represents 'perishable but regenerative life' (1999:74). That is, the serpent symbolises the Paiela philosophy that to perpetuate life, one must expend life. While I whole-heartedly agree with her assertion that Huli and Paiela people tend to conceptualise reproduction and descent as 'serial regeneration' of the previous generation, I do not think that Tai Yundiga, or snake imagery in general, is an instance of this conceptualisation. On the contrary, for the Huli, serpents represent freedom from this cycle of life, reproductive expenditure of life, death, and continued life in the next generation. Self-renewal through sloughing off the used-up body precludes the possibility of decrepitude and the necessity for reproduction. Indeed, Biersack's own data suggests that snakes represent this freedom from the human predicament of aging and loss of vitality. As one of her informants says in his description of heaven, 'everything will be new, and *our skin will be changed*. Whatever had not worked on earth ... will work in heaven. It will be just like changing a tire: what was bad will be replaced ... and we won't die' (p. 79, my emphasis). It is this mode of being, what Biersack calls a 'postbiological order,' that serpents represent for the Huli. Moreover, among the Huli, Tai Yundiga is far from dead in that he is able to enter into secret alliances with humans.

15. The good company has come and called out to you, 'Mother of Life', but no, you only say 'I am the Mother of Death.' You should cooperate with the company but you don't: '*I Company payaleme ibuwa "Habe Ainya-o" laragola, ndo. O piago warabe lowa "I Homabe Ainya peto" laregoni. Company la piabe paya hangu I pialu ha nabi.*'

16. Also see Lattas 1992:31 for another example in myth of colonised people imagining themselves as unworthy of wealth because of being '*bikhet*' – Tok Pisin for wilful, arrogant, and recalcitrant.

17. In the traditional myth, as well as in the current parable, no reason is given for why the female character responds with silence when called 'Mother of Life.' Interestingly, this sort of conversational pattern frequently occurs between husbands and wives. While husbands do not go around

calling their wives 'mother of life,' they do when approaching their wives' houses call out 'mother of (their child's name).' Often, if women are annoyed with their husbands, they will simply not answer, which invariably makes their husbands quite angry. This marital discursive pattern of address, followed by sullen silence, followed by sarcastic anger is quite common, and women can easily explain why they refused to answer. An important politics of power goes on in these brief exchanges, with women's silence expressing a resistance to being ordered about. Significantly, this discursive politics of gender is left out of the myth, making the female figure seem irrational and even malevolent.

18. I use the term fantasy deliberately to invoke notions of desire and wish-fulfilment. While I do not do it here, I think it is possible to make the argument that the disruption of traditional ways of being and the alienation from the land caused by mining could cause the sort of trauma that would re-engage primary psychic tensions. In the case of mining, for example, anxieties about disempowerment and about disrupted attachments to the land may be played out using images – such as snakes and faeces – from the unconscious.

4

WHO AND WHAT IS A LANDOWNER? MYTHOLOGY AND MARKING THE GROUND IN A PAPUA NEW GUINEA MINING PROJECT

Dan Jorgensen

The problem I want to discuss in this chapter concerns alternative ways in which indigenous people attempt to formulate their relation to land in the context of contemporary mining developments in Papua New Guinea (PNG). One important aspect of my argument will be that the status of landowner in PNG is, despite lawyerly desires for clarity, uncertain at best. The characteristic landholding situation in PNG is one in which a range of overlapping rights may apply to any given stretch of territory, and numerous candidates may justifiably argue their claims in any given case, a state of affairs complicated by the fact that mining has few analogues in traditional practice. I illustrate some of these issues with an example of the ways in which Telefolmin have dealt with land matters arising from the proposed Nena mining project. The Nena case poses additional problems because the mineral deposits in question are located in an area with a turbulent history of warfare, expansion, and displacement. The proposed mine site is thus situated in a contested field in which the struggle for control over territory has generated alliances and enmities for at least 150 years, and one part of my argument is that this history shapes rival claims to mining benefits and is itself subject to formulation in the light of current interests. Finally, I argue further that the involvement of the state in the recognition of land rights introduces an added dimension to land issues that is inevitably political in nature, making the ability of local people to exert varying kinds of influence or pressure on the state an important component of the de facto definition of who and what a landowner is.

Post-independence PNG has undergone a growing involvement in large-scale mineral exploitation as a core strategy of national planning. Despite an early commitment to decentralised rural development, smallholder production was inadequate to the task of financing government operations, and returns from the Bougainville copper project held out the prospect of mining as an important source of revenue. In the 1980s and early 1990s, PNG enjoyed a mining boom with the inauguration of a number of mines, including Ok Tedi, Porgera, Misima, and Mount Kare (Connell and Howitt 1991, Jackson 1982, Polier 1996). Yet even as national revenues rose, problems associated with mining multiplied, and landowner dissatisfaction quickly became a central issue facing the national government. Although the Bougainville rebellion is the most spectacular instance of landowner discontent (May and Spriggs 1990), it is by no means the only one. The Ok Tedi mine has been plagued since its inception by strife over ecological damage to the Fly River system, while Porgera and Mount Kare have been the focus of intense and persistent wrangling between local groups and competing jurisdictions (Vail 1995). Not surprisingly, the national government has made several changes to the terms of mining agreements and the ways in which they are negotiated in an effort to forestall further problems (Jackson 1989). Although the specifics vary from case to case, the main emphasis has been on the balance between the benefits local people receive in relation to the social and ecological costs mining operations incur, one result of which has been a steady rise in the size of benefits packages offered to landowners.

Whether such efforts will prove successful is not yet clear. Mining agreements are easier to make than to sustain, and some observers are pessimistic about the possibility that any agreement will live out its appointed lifespan (Connell and Howitt 1991; Filer 1990, 1992). Quite apart from problems associated with compensation for damage or the shares local people receive relative to various levels of government, however, each mining project generates additional problems that arise from local dilemmas of distribution (Filer 1997b).[1] Here the chief difficulty is to determine how widely or narrowly to define the pool of recipients of the benefits package, which may include not only royalty payments and occupation fees, but also housing, health services, training, employment and assistance with business development. In general, the larger the pool, the smaller the shares in the overall package. A few recipients may enjoy large shares in the benefits, but at the cost of aggravating problems associated with uneven distribution within the project area. If, on the other hand, the net is cast widely to include as many beneficiaries as possible,

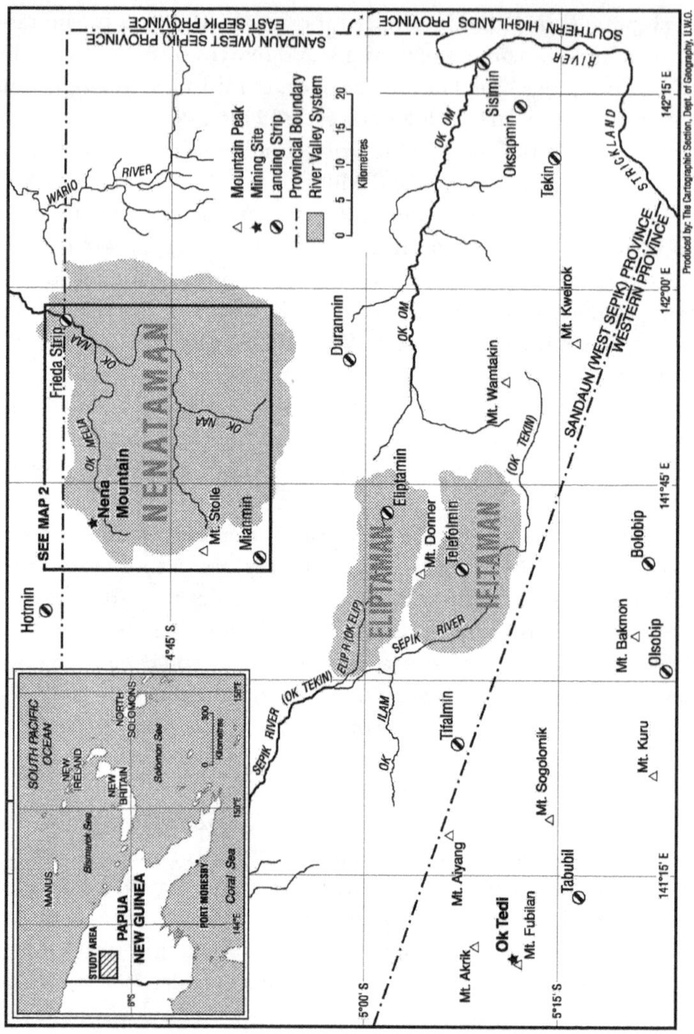

Map 3: The Ok Tedi mine and neighbouring Ok region.

the risk is that those who feel they have the strongest claims may find their shares diluted below a level they find acceptable. Virtually all of these problems turn on whether and how various claims over land are to be recognised, issues that are tied up with determining who is and is not a landowner.

Telefolmin and mining

For the Telefolmin of Sandaun Province, these questions have assumed growing importance as the number of mining projects in their environs has grown over the years. Numbering about 4000, the Telefolmin live in the mountain valleys of the Sepik headwaters, an area to which geologists were attracted as early as the 1930s (Kienzle and Campbell 1938), well before the establishment of colonial administration. In the 1960s and 1970s, mineral exploration began in earnest. To the north, the Frieda River prospect (which includes the area known as Nenataman) was explored by Carpentaria Exploration (CEC); to the west, the Tifalmin and Mount Fubilan prospects were explored by Kennecott Copper. Although Kennecott abandoned their prospects as a result of failure to reach agreement with the government over royalty and taxation issues (Jackson 1982), Fubilan was later developed by a consortium which included the PNG state. This became the Ok Tedi mine (see Map 3), construction of which began in 1981, a year which marks a dramatic increase in mining's importance in PNG's affairs.

The idea of mining is thus nothing new for Telefolmin: the better part of a generation has grown up amid talk of mining, and most adult men have been employed in the mining industry in one way or another. In the 1970s, when the chief sources of income were labour contracts on distant plantations, working for mineral survey teams provided much more lucrative employment and had the added attraction of being close to home. Quite apart from the cash such employment provided, working for mining companies was always tied to the promise of better things to come: in an area that was a remote backwater, it defined the horizon of expectations for the future. Mining thus became synonymous with the *developmen* that people hoped would redress the obvious disparities between Telefolmin and less marginal parts of the country, transforming them from 'bush kanakas' to people with money, just like people in the towns where they worked.

When PNG gained independence in 1975, one of the chief worries local people had was whether or not mining at Fubilan would go ahead, and news of disagreement between Kennecott and the government was met with resentment and the fear that the project would, like so many other hoped-for opportunities, come to nothing. When the Ok Tedi project actually commenced, however,

Telefol expectations were disappointed. Although the area enjoyed unprecedented prosperity as a result of the wages brought in by workers, the Telefol role was largely confined to providing labour at the lowest rungs of the camp hierarchy.[2] Peak employment took place during the construction phase, with as much as 40 per cent of the adult male population working at the mine or its townsite of Tabubil. Wages were high by PNG standards, but the conclusion of construction a few years into the project reduced the overall number of jobs available to local men. Apart from the issue of employment, however, Telefol dissatisfaction centred on the fact that they were not recognised as landowners of the mine site. In order to understand what this means, it will be necessary to turn to a discussion of Telefol perceptions of their place in a regional system of relations.

In common with their Mountain Ok neighbours (Barth 1971, 1987; Craig and Hyndman 1990; Gardner 1981, 1983; Hyndman and Morren 1990; Jorgensen 1996, Morren 1986*a*, 1991*a*, 1991*b*), Telefolmin identify themselves as 'Min', a category embracing those peoples who trace descent from an ancestress known as Afek (see Jorgensen 1996). Afek is said to have travelled throughout the Sepik-Fly headwaters and established local populations and their customs along the way. At the end of her journeys she came to a site overlooking the upper Sepik, where she is said to have built a large spirit house she named the Telefolip (Jorgensen 1990*a*). This house became the nucleus of the village of Telefolip, from which the Telefolmin take their name. Telefolip was at the centre of a regional male initiation cult and was the site of rituals which underwrote taro fertility for the entire Min region. From this location Telefolmin expanded over the span of perhaps 200 years to their present distribution in the upper Sepik, Elip and Frieda valley systems, known locally as Ifitaman, Eliptaman, and Nenataman.

Relations between Telefolmin and their neighbours were varied, ranging from bitter warfare to cordial trading, but the Telefol custodianship of Afek's house gave them ritual pre-eminence throughout the region, a position acknowledged by allies and enemies alike. Telefolmin were thus part of a regional system whose precolonial basis was rooted in mythology, ritual, and a perception of linguistic and cultural commonalities among those peoples claiming descent from Afek.[3] When the Australian administration established its first patrol post in the area at Telefolmin following World War II, the location of the centre of colonial authority seemed to reaffirm this configuration of relations.[4]

The announcement that minerals had been discovered at Fubilan was greeted with enthusiasm, a reaction that owed some of its force to the peculiarities of the mine site's location. Situated almost due west of Telefolmin, and near the Wopkaimin hamlet of Bultem, Fubilan enjoys a special place in Telefol mythology as the site of the Land of the Dead (Bagelam) established by Afek. When Afek killed her younger brother, she sent him along an underground track to Fubilan, where he made a place for the dead and created the source of shells (*bonang*) and the stone *fubi* adzes (hence Fubilan) which were the principal items of traditional wealth. Thus the mythology of Afek not only linked various peoples in a genealogy of her descendants: it also mapped relations among ancestors, the living, and wealth at Fubilan. Given the ritual primacy of Telefolip and the role of Telefolmin as custodians of Afek's legacy, the government's decision to recognise the Wopkaimin as Fubilan's landowners generated consternation and resentment.[5] This reaction was intensified when it was rumoured that all additional benefits from the mine would be allocated to the Western Provincial Government, thus excluding Telefolmin and most other Min peoples as well (Map 3). Dissatisfaction over Ok Tedi thus added to long-standing discontent over the division of Min people into two provinces and gave new impetus to efforts to establish a 'Min Province'. Such a province would consolidate the territories inhabited by Afek's descendants, with the Ok Tedi mine as its centrepiece.[6] The movement to establish a Min Province played out in various arenas during the early 1980s. At home in Telefolmin the mythology of Fubilan took on added significance as Ok Tedi went ahead. As the mythological source of wealth, Fubilan was readily understood to be the source of gold that in some way became money. The onset of mining operations coincided with the lapse of trading links between the newly wealthy Wopkaimin and Telefolmin, with the result that traditional valuables were no longer available. Telefolmin linked this to the physical dismantling of Fubilan, and argued that some form of compensation should be forthcoming for this loss. Further afield, Telefol politicians at both provincial and national levels pressed the government for some sort of recognition of a special status for Min people *vis-à-vis* the mine, regardless of which side of the provincial boundary they lived on.[7]

In the event, these efforts bore fruit with the signing of the Telefomin District Development Agreement (TDDA) in 1983. The agreement acknowledged a special relationship between the peoples of Telefomin District (that is, Min) and the Ok Tedi project, and called for the establishment of a national level high school at Telefolmin and the creation of an annual fund for development projects within the District. Although the aim of creating a Min Province was, for the time being, not achieved, the TDDA amounted to official recognition of at least some Telefol claims, a recognition that brought with it tangible benefits stemming from the Ok Tedi project.

This is far from the whole story, however, for the establishment of the TDDA was given a local reading that viewed the agreement as a token of the government's acceptance of the substance of the mythology linking Afek and Fubilan. In Telefol eyes this came to more than the recognition of vague 'ethnic' or 'traditional' ties to the mine site and implied agreement with one of the myth's central contentions, namely that *the source of the wealth* (in this case, *fubi* adzes) *was created by Telefol ancestors* (cf. Trigger and Robinson, this vol.). This point was made graphically during the signing ceremonies in Telefolmin, when PNG's governor-general was publicly presented with a *fubi* adze by the councillor of Telefolip village (see Jorgensen 1990b).

The Nena context

Although Ok Tedi is the most significant mining project on the Telefol horizon, the Frieda prospect to the north of Telefolmin also figures prominently in local perceptions. The valley known as Nenataman, in which the Frieda prospect is situated, is home to roughly 600 Telefolmin in the villages of Ok Isai and Wabia. Their closest neighbours and former enemies are the Paiyamo of Bapi village to the north, and the Miyanmin of Wameimin on the western margins of the valley (see Map 4).

Mineral exploration in Nenataman offered intermittent employment since the 1960s, albeit on a much more modest scale than Ok Tedi. Until the late 1980s, however, there were few indications that operations were likely to go beyond the exploratory stage. This changed when Highlands Gold Ltd. (HGL) became the latest of a succession of developers to take an interest in the area. HGL began new exploration in the upper Frieda system and located previously unknown copper deposits along the Telefol-Miyan frontier at Nena

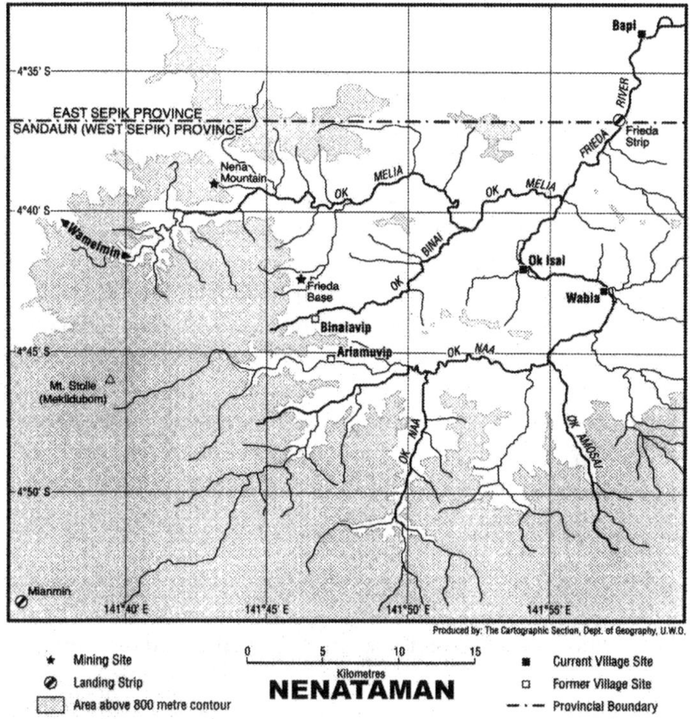

Map 4: Nenataman

Mountain. This discovery revived the possibility of mining in the area, which local people welcomed: Telefolmin living in the vicinity held nightly prayer meetings in the hopes that mining would finally go ahead. By 1992, they seemed to have received their wish with the announcement that the prospects for the Nena site were good.

Not surprisingly, however, this news brought with it a series of questions about land ownership and royalties that dominated local concerns, one expression of which was the formation of the Frieda Mine Landowners' Association (FMLA) in 1992. These concerns prompted the FMLA to invite myself, George Morren and Rune Paulsen to conduct a genealogical study under the auspices of HGL in 1995. Don Gardner, who had already been working independently as a consultant on social impact issues, constituted a fourth member of our team. Our brief was to compile genealogical and census information, as well as to record local practices concerning land and land claims.

One of the complexities surrounding mining prospects in Nenataman is that the region as a whole has been the site of intense and prolonged warfare between a number of neighbouring groups. The most prominent line of cleavage is that between Telefolmin and Miyanmin, both of whom assert claims over Nena Mountain. But Nenataman has also witnessed a succession of other occupants who have been annihilated, absorbed, or displaced. It is against this history that claims to land – and potential royalties – must be seen.

As far as it is possible to determine, the inhabitants of Nenataman in the early 19th century comprised a number of groups, including the Hendari, Nambulu, Ulumalo and Untou, none of which are now extant. Of these, the Untou are regarded as the most important, and it is likely that the other names refer to subgroupings within this larger category. Neighbours of the Untou included the Miyanmin to the west, the Iwam, Owininga and Paiyamo peoples to the north, and the Wario and Duranmin peoples to the east. To the south were the Iligimin, a Telefol-speaking group inhabiting the valley known as Eliptaman. By mid-century the Untou were being pressed by the Iligimin, who mounted a series of attacks which wiped out a number of Untou settlements and drove others to their neighbours for refuge. The Iligimin established the villages of [Old] Wabia and Ariamuvip in Nenataman, from which they continued their raids. Their position, however, was not consolidated before they found themselves embroiled in a new round of fighting with their neighbours to the south, the Telefolmin of the upper Sepik valley (Ifitaman). In the course of this conflict the Iligimin burned the central spirit house at Telefolip, and Telefolmin in their turn mobilised allies from Urapmin and Faiwolmin in a campaign to annihilate the Iligimin. After a sustained and bloody series of fights, the Telefolmin and their allies destroyed all of the Iligimin settlements in Eliptaman. Most of the Iligimin were slaughtered, but young women, children, and selected men were spared and incorporated into Telefol villages.

After a brief hiatus, Telefolmin and incorporated Iligimin resettled Eliptaman, and from this base they attacked and then absorbed the Iligimin settlements in Nenataman. Once established in Nenataman, Telefolmin took over where the Iligimin had left off and carried out a systematic series of raids aimed at clearing the valley of its other inhabitants. The combined pressure of Telefol raids, epidemics, and occasional raids by Miyanmin completed the destruction of the Untou as a group around the turn of the century. Fighting in the western part of Nenataman continued, however, for Telefolmin and Miyanmin had become embittered enemies. To the north and east, Telefolmin fought intermittently with the Paiyamo, Wario, and Duranmin peoples, a situation that prevailed until the end of the 1950s. Raiding came to a close in the early 1960s, when a Telefol raiding party with captives in tow stumbled across a geological survey team at the Amosai-Frieda junction. Fearing they had encountered a government patrol, the raiders returned home feeling that further forays had become too risky.

Social organisation and customary land tenure

The basis of Telefol social organisation is the village (*abiip*), usually consisting of about 200 to 300 people related by cognation and affinity. Telefol society lacks the elaborate forms of ceremonial exchange common among the Highlands societies east of the Strickland, and inter-village relations tended instead to be mediated through an elaborate system of male initiations centred on the ancestral village of Telefolip. Since the late 1970s most Telefolmin, like their neighbours, have become Christians (Jorgensen 1981b, Robbins 1995), and churches have since that time served as an important channel for relations between different villages.[8]

The Telefol preference is for village endogamy, with marriage by sister exchange and payment of bridewealth. This ideal, however, was not always realised in practice, and in the days of warfare it was common, particularly in Nenataman, for Telefol warriors to abduct women and children who would be incorporated into their own families, a practice which has some implications for contemporary claims.[9] Many of those who are descended from incorporated war captives now maintain relations with kin in groups that were formerly enemies of the Telefolmin. The kinship system is fully cognatic, with non-exogamous descent categories known as *tenum miit*. *Tenum miit* have no corporate functions and are not themselves landholding units. Children are affiliated with the *tenum miit* of both parents, and it is common for individuals to claim 'membership' in as many *tenum miit* as their ancestry warrants. Although the majority of Telefolmin in Nenataman belong to the Kayalikmin *tenum miit*, not all do so, nor does Kayalikmin identity preclude other identities.

The Telefol pattern of land use is extensive, with large areas of land being utilised for a diverse range of activities. Taro gardening is practised on the basis of shifting cultivation, requiring large reserves of secondary fallow. Hunting and sago exploitation in Nenataman are also important, but result in minimal disturbance of the landscape, and these factors conspire to produce a situation which gives the superficial impression of large tracts of apparently empty or under-utilised bush. In fact, as elsewhere in PNG, virtually all land is subject to claims of one kind or another.

Telefolmin recognise a wide range of rights in land on the basis of a limited number of principles. While the principles themselves are simple when taken singly, their realisation in practice can be complex. Practices governing settlement rights differ from those concerned with gardening, which in turn are different from those affecting hunting or sago. The interplay between various kinds of rights and claims can be a difficult matter, especially when different ranges of people claim these rights. Hunting and collecting rights are diffuse and apply to bush throughout the territory associated with a village or clusters of villages. Land for taro gardening is governed by a more restrictive set of conventions organised around the principles of first clearance, bilateral inheritance or permissive usufruct. Because Telefol agriculture puts a premium on cultivation in different altitudinal zones, most people have claims in several different locations. Both men and women have independent rights in named tracts of land, which they may pass on to descendants of either sex.[10]

Ecologically, Nenataman is a low-lying area with an extensive cover of primary forest, in contrast with the upland valleys of Eliptaman and Ifitaman[11]. Game – particularly cassowary and wild pigs – is abundant here, and the low elevation and high rainfall are particularly conducive to sago, which is absent from the higher valleys. Most sago grows in swampy stands which were established by the original inhabitants or by pioneering ancestors in the previous century. Sago trees are owned by individuals or siblings, although people are normally generous in allowing others to make use of their trees. Sago is used in a number of ways, but is most important as an adjunct to hunting and, in the past, to warfare.[12] Groups of warriors would plant sago, *marita* and arrow cane (*biil*) at bivouac sites far from home in order to provision future expeditions with food and weapons, and in this way a far-flung network of sago stands was established along routes followed in raids. Although warfare is a thing of the past, hunters still fell sago logs to lure wild pigs, which may be more easily ambushed in this way.

While the principles of land tenure are straightforward enough, they may be conditioned and qualified by other factors. An important premise of Telefol landholding is that one must exercise land rights in order to maintain them. A claim to land – particularly garden land – must to some extent rest on a history of work and care which leaves one's 'mark' on the land. Such marks take various forms, the most persuasive of which is the maintenance of second-growth forest by repeated clearance, but may also include plantings of *marita* pandanus or fallow trees such as casuarina. Conscientious parents make an effort to 'show' their children all the land in which they have claims, and make a point of returning to previously cultivated tracts in order to keep their claims fresh. Because sago requires periodic tending to be productive, it is possible for individuals to lodge claims to sago that they have repeatedly tended, particularly if the nominal owner has been lax or unable to look after his trees.

In general, the Telefol system is flexible but contains a number of areas of ambiguity in which trouble could arise, particularly since the pattern of bilateral inheritance guarantees a large number of overlapping claims. Despite this, disputes over gardening land are surprisingly rare, even in Ifitaman, the most densely populated part of Telefol territory. This is in part because there was always enough garden land to go around, and partly because disputes between fellow villagers are frowned on in Telefol culture. Where two rival claimants cannot reach accommodation regarding a particular tract, the land is likely to remain uncultivated because each is believed to have the power to render the land infertile. As with garden land, disputes over sago are rare because it is believed that individuals can easily 'spoil' the sago so that no one benefits. Such considerations serve as a powerful deterrent to the unbridled pursuit of claims against fellow villagers. At the same time, however, they also indicate a complementary readiness to destroy the bone of contention in order to frustrate an opponent.[13]

Multiple claims in Nenataman

Landholding at the collective level is best understood as a matter of politics rather than economics, of sovereignty rather than property. A community may hold land either by ancestral settlement or by conquest, and each village or cluster of villages claims a large zone for the use of its members. All rights are subject to the proviso that one is able to enforce them, and this has meant that effective land rights are ultimately based on membership in a political community capable of defending such claims against outsiders. Telefolmin hold land in Nenataman by conquest, but this general characterisation conceals a range of complexities, beginning with the definition of the political community in which collective rights are vested. Whereas pioneering may be undertaken on an individual basis, conquest is always the outcome of group effort, and in the case of Nenataman, the relevant group has had various configurations over time, depending on the phase of settlement and the degree of opposition encountered.

In the early phases of settlement following the Iligimin war, this community consisted of the combined population of Telefol settlements in Eliptaman, with some elements from Ifitaman. These villages were the source of the Nenataman colonists and supported them in warfare. When Telefol settlement in Nenataman was relatively well established near the beginning of this century, the villages of Binaiavip and Ariamuvip functioned more or less independently of Telefolmin elsewhere.[14] A couple of decades later, however, when the Miyanmin staged a raid on Binaiavip, links to Eliptaman and Ifitaman were once again activated to mount a massive retaliatory raid. Similar assistance was provided in fighting against the Wario and Duranmin peoples to the east. This history of support in warfare provides the basis for strongly argued residual claims by the descendants of fighters who came to the assistance of Binaiavip and Ariamuvip, even though they themselves did not settle there.

An additional complication arises from the assimilation of numbers of captives taken in warfare. Some of these individuals and their descendants seem to play a special role in Telefol claims in Nenataman by serving as 'sponsors' for Telefol occupation. The clearest example of this is the case of Obulum, an Untou man who was abducted as an infant in a raid along the Melia. After being reared in Ifitaman and Eliptaman, he was brought back to Nenataman and 'shown his land' by the man who had abducted him in the first place. Obulum settled in Nenataman and had a number of descendants, and while most people assert general claims on the basis of conquest, others are equally comfortable asserting rights deriving from the original inhabitants.[15]

This overall picture of land rights in Nenataman must be modified to take various claims by non-residents into account. The Telefol preference is for village endogamy, so husbands and wives are usually members of the same community. In cases where husband and wife originate in different communities, however, it is possible for family members to take up rights in both communities. This can lead to a situation of shifting residence as family members move from one community to the other in the course of working their land, and the descendants of such unions may claim such rights providing they have actually gardened the land in question. In the longer term, this means that claims over community resources may be entertained by people who reside elsewhere.

In addition to claims arising through intermarriage, various kinds of rights may be extended to outsiders on the basis of friendship or kinship. Nenataman's attraction for Telefolmin has been mainly in terms of its wild resources, and local people have become accustomed to hosting upland visitors in search of forest products absent from their home valleys (see Morren 1979 for a Miyanmin example). Sometimes people invite neighbours to join them in clearing and cultivating a tract, and such invitations may also be extended to people from other villages who happen to be visiting when a new garden is being made. A gardener will occasionally permit a man to plant *marita* in his or her garden, the fruits of which will be shared on the basis of the labour invested in tending the tree. Further, outsiders may also exercise rights over sago. The fact that sago was often planted in association with warfare and hunting means that some of those who participated in such expeditions (and their heirs) have rights over such trees. These rights are conditional on maintenance, and those who fail to properly tend their trees in effect lose the force of their claims. At the same time, as mentioned above, those who invest the effort in looking after sago (or *marita*) thereby establish a claim of their own. Such claims have proliferated in the wake of mineral exploration at Frieda Base, when a number of men employed there looked after nearby stands of sago. As a result, it is not uncommon to hear that word has been sent to a friend or relative in Eliptaman or Ifitaman to come and harvest their sago or *marita* in Nenataman. All of these contexts involve an engagement in the care and working of the land, and if a person has been more assiduous in working a tract than its nominal owner, his or her claims may come to rival or supersede those of the latter, and it sometimes appears that the rules of inheritance turn on descent from the most recent cultivator, regardless of the nature of the latter's original claim (cf. Welsch 1991). Under these circumstances, the distinction between permanent versus temporary rights tends to become blurred.

'Prehistory' and 'true history' in Nenataman

This account of Telefol land customs provides a basic framework for discussing the identification of landowners, but it misses the liveliness with which local people advance arguments to support their claims in the charged atmosphere surrounding the Nena project. This is so not only because the perceived stakes are so high, but also because uncertainties surrounding the definition of landowners fuel worries about losing out. These uncertainties arise from several sources, not the least of which is the fact that customary practices offer only ambiguous guidance on who counts as a landowner for the purposes of allocating mining royalties. Such difficulties are accentuated by the fact that Nenataman Telefolmin are not the only ones who assert ownership over the territory in which the Nena project is situated. Early explorations of mineral deposits in Nenataman focused on an area located near the Frieda Base in the headwaters of the Ok Binai. This is an area where Telefol claims are relatively undisputed, but the richer deposits at Nena Mountain are located along the Telefol-Miyan frontier, and neighbouring Miyanmin have been active in pressing their case. Thus the welcome news that the Nena finds made a mine feasible simultaneously called previously secure claims into question.[16]

The anxieties this generated were palpable during the course of my fieldwork, when my arrival in a Telefol village would often be greeted by exclamations of '*Dan i kam – mipela win nau!*' ('Dan has come – now we'll win!').[17] My arrival had been anticipated, and people came forward with a variety of different kinds of evidence, or *wisnes* ('witnesses') prepared. Most of this came in the form of oral accounts of personal genealogies and land use histories, with detailed listings of the marks they or their ancestors had left on the ground around the proposed mine site. This testimony was backed up with the aid of locally recognised collateral evidence in the form of ancestral relics, war trophies, and the texts of commemorative songs that provide a crucial medium of Telefol oral history. Some came with typescript statements or computer-generated census lists that made an excellent approximation of official village registers. Others who were familiar with government notions of landholding began talking of traditional cognatic descent categories (*tenum miit*) as 'clans', complete with patrilineal descent.

When confronted with the claims of their Miyan rivals, Nenataman Telefolmin point to particular sites in the vicinity of Nena Mountain where they have left their mark and refer to the history of fighting in the area. After the Miyan raid on Binaiavip earlier this century, Telefolmin mounted a massive attack on Miyan settlements at the Usake and May River junction with the help of a large contingent of warriors from Eliptaman and Ifitaman. Although Telefolmin subsequently relocated to more easterly sites in Nenataman, this raid went unanswered, and Telefolmin argue that this silence represented Miyan acquiescence to Telefol control over the western portions of the valley.[18]

Despite the importance attached to narratives linking people, place and activity, however, many prefer to base their case on arguments couched in the register of myth. In the words of a member of the FMLA executive, the first sort of account constitutes the 'prehistory' of the area. The second, however, is claimed to be more important, and amounts to the 'true history' of Nenataman.[19] It is to this latter that I now turn.

A mine and two myths[20]

Shortly after I arrived in Nenataman, I was taken aside by a deputation of senior men from Ok Isai, including one of the local pastors. They proceeded to tell me the story of the origin of *bangelii*, small stone adzes formerly used by women to clear garden undergrowth. This myth was repeated to me with variations throughout Nenataman and in Telefol settlements elsewhere, and was said to be the real explanation for the existence of minerals in Nenataman, as well as for the Telefol occupation of the valley. I reproduce three versions of the myth below.

The origin of bangelii, *version 1*

There were three big ancestors at Telefolip: Afek, her younger brother Umoim, and Kwiinagim. When Afek was at Telefolip they once had an initiation to which a tall white man with long hair came. He wanted to participate, but the others looked at him and thought, 'he's by himself – why not kill him?' So they held him and tied him with thorny rattan. After they bound his arms and legs, they cut his achilles tendon and he died. Then they cut him up and took his bones. They put him in a cave at Mofumkot [near Telefolip]. He was there for three days, and there was thunder, lightning, rain and earthquakes. Two women – a Wopkai woman and a Sibia woman – cut some sweet-smelling wood and brought it to the cave. But when they moved aside the stone that covered the entrance and looked inside, they found that he was gone. People told the women that he had gone to Telefolip, so they took the wood with them to Telefolip, but he wasn't there either. He had crossed the Ifi and gone to Wimtem [near Urapmin], looking for Afek. He went to cross on the bridge over the Sepik, but Afek had cut its supports and he fell into the water when he got halfway across. He became the first human to swim when he fell in the water, and emerged safely downstream. In the meantime, Afek was on the south bank of the river, busy making things like machines and cars. He sneaked up on her from behind and grabbed her, but she took a seven-pound hammer and swung it, crying, 'Kwiinmimi!' Killing him with one blow, she sent him back to Telefolip and told him he would own the fork in the road followed by the dead on their way from the land of the living. He returned to Telefolip for a second time and placed a *fubi* adze at the entrance to the road to the Land of the Dead [in Telefolip], and he placed another underneath the Fubilan [Ok Tedi] mine. At Nena he put the *bangelii* adze in the ground – Umoim came and hid it here. Then he went underground and returned to Telefolip.

Our ancestors wanted to follow this and came looking for this *bangelii*, and that's why they came here [to Nenataman]. The first Kayalikmin ancestor was a dog named Titak-unam, whose wife was the *Kayal* bandicoot. Their son Wipnagim married Kunkomen; they are the guardians of the *bangelii*. Their child was Kalasim, who was Oyapnok's father. Oyapnok was the Kayalikmin commander [*komanda*] – he stayed at Telefolip and sent off soldiers [*soldya*] to the Fak and the Elip – Kotremengim, Binimnak, Alisep, Maisep. They found the *bangelii* here [in Nenataman], and when they found the Paiyamo, Untoumin, Akiapmin and Miyanmin here, they chased them out.

Umoim's netbag[21] contains his bones and a shovel, a spoon, cloth, and a plate, but they [the people of Telefolip] have hidden all these things. He's used for taro ritual and war ritual. He opened the road for the dead – he can open the door. Telefolip has two roads, a good one and a bad one, and he watches over the fork where the two divide. Wipnagim is Umoim's second name. The reason the Kayalikmin came and took Nenataman was that they wanted the *bangelii* – the dog hid it and the Kayalikmin wanted to get it back. That's why they hunted the enemy.

The *bangelii* was used for garden work and in bridewealth. *Fubi* too. It has meaning and power, and this is why Kayalikmin grabbed Frieda. And at Amosaitaman there's *tong* [fire-making pyrites] that was put there at the same time. Miyamduvip, Komduvip and Ankem[22] should all get pay as Kayalikmin places. Aisep, Maisep and Obulum were Atemkayakmin and were like Oyapnok's 'slaves' [*slev*]. Oyapnok sent people to the Elip first, then to the Fak, then to the Henumai, and each time he said 'go further'. When they came to Frieda, he said, 'okay – Kayalikmin dog-children are strong for the fight – chase the enemy away, surround those people and kill them.'

The origin of bangelii, *version 2*

At Telefolip there was the old woman Afek[23] and Kwiinagim. Kwiinagim's son was Titak-unam; he was a dog and they have kept his head at Telefolip. His son was Tumunim and Tumunim's son was Yaapnok, whose bones are also held at Telefolip. Yaapnok had three sons, Alofengim, Yasipnok, and Maisep. Kwiinagim had three things: the *fubi* adze, which he buried at Tabubil; the *mook* adze, which he hid at Porgera, and the *bangelii* adze, which is hidden at Nena. The dog hid these things. And the dog [that is, the Kayalikmin ancestor] can smell these things. When the Kayalikmin came here fighting against all sorts of people – Falamin, Kasangkelmin, Oksapmin, Duranmin, Wario, Miyanmin – all these people said, 'we didn't take something of yours [that you should come fight us]!' But this stone adze is what the fighting was about. The Kayalikmin ancestor came, smelled it, looked around for it and made a camp on top of the gold.

The origin of bangelii, *version 3*

At Telefolip there's a spot where John the Baptist baptised Jesus, and the tree upon which Jesus was crucified is there as well.[24] After Jesus was killed, they put his body in Nangalamtem [a sacred cave nearby].[25] When I went to Telefolip, I asked a man there about this, but he shushed me, saying that if people heard us talking about this, they

> would kill us. This is talk about gold. There are four [stone] things that were given to our ancestors, and they are like a man. At the head is the *fubi* adze, which bosses the gold at Ok Tedi; one leg is the *mook* adze, which is at Porgera; the other leg is the *bangelii* adze, which is here at Nena; the heart is the *tingii* [stone club head] at Telefolip. White men came and were looking for gold and found three of these, but they didn't find the *tingii* at Telefolip – they hid it well there. The *tingii* covers the entrance to the Land of the Dead there.

The three durable items of traditional Telefol life – human bones, shell valuables, and stone tools – are all invested with powerful symbolic valences associated with life-giving powers (Jorgensen 1983, 1985), and many of these find resonances in contemporary myths linking stone adzes with mineral wealth.[26] Stone adzes were essential to traditional subsistence and were both sources of wealth (for example, pigs, foodstuffs) and wealth items in their own right. Originating outside Ifitaman, stone tools were understood to have a subterranean source and were mythologically associated with the dead. Local people now make a general association between geologically anomalous locations (marked by discoloured streams, different forms of rock, stunted flora, and so on) and the presence of valuable minerals, and many such locations were already viewed as sacred or powerful sites (*amemtem*), of which Nena Mountain is one. More generally, sources of stone blades are equated today with sources of gold, and this achieves a tolerably good fit with experience in the light of the fact that Umoim's Land of the Dead and the traditional source of *fubi* adzes (Mount Fubilan) is the site of the Ok Tedi mine (Jorgensen 1990*b*; see above). The myth of the origin of *bangelii* draws upon a range of Christian and traditional prototypes. Here three traditional episodes are compressed into elements of a single story: the killing of a tall, light-skinned stranger at an initiation, the story of the death of Umoim, and the story of Kwiinagim's attack on Afek. The killing of the light-skinned stranger is derived from a story about the killing of the son of the Sun (Ataanim)[27] at an initiation; other versions identify the victim as the ancestor of Europeans and, even more recently, Jesus Christ. The story of the crucifixion and resurrection are also evident in the handling of the protagonist's [first] death, in which the roles of Mary and Mary Magdalen are played by 'a Wopkai woman and a Sibia woman.' These identifications are crucial, for they evoke the Ok Tedi and

Nena mines: Wopkaimin are located around Ok Tedi, and the mountain known as Sibia-dubom [the Knob] overlooks the Nena mine site.[28]

Umoim, Afek's younger brother, is implicitly identified with Jesus Christ in the first and third versions, evoking a widely known myth about the origin of Europeans which accounts for the relative wealth of Europeans and the poverty of Telefolmin. This episode is linked to one traditionally told about Kwiinagim, in which he follows Afek and rapes her near Urapmin. In the Kwiinagim tale, he comes upon Afek as she is changing the skins of insects, snakes and lizards, and the point of the episode is to explain that his intervention prevented humans from having the ability to renew their skins when they became old, an intervention that costs him his life. In the present version, this episode follows a similar logic, but implies that Umoim's intervention resulted in the loss of the ability to make machines and cars.[29]

The remainder of the myth links the origin of *bangelii* to the Kayalikmin-led Telefol invasion of Nenataman. Here the point is that the *bangelii* was buried under the ground the way a dog buries a bone, and the subsequent Telefol incursions into the north are to be understood as attempts to retrieve an ancestral legacy, just as a dog digs up a bone once buried. The adze blades are identified with body parts, and they were secretly held to be bones, which was why a dog buried them and why a dog could smell them. The myth also makes the point that Kayalikmin aggression is linked to the aggressive nature of their dog-ancestor.

The myth of the *bangelii*, like most other Telefol myths, sets Telefolip as the original scene of the action and accounts for what many see as a puzzle: that goldmines are found in places selected by Telefol ancestors, but that no gold in Telefolmin proper (let alone Telefolip) had been discovered. This is because the seniors at Telefolip had concealed this from prying European eyes in order to safeguard the entrance to the Land of the Dead. The *tingii* at Telefolip is of special significance, for it is at the heart of the gold, and its shape – a perforated stone disc – is held to be the prototype and source[30] of PNG's one-kina coin, also a perforated disc.

While the details vary in individual accounts, the gist of the myth is widely repeated throughout Nenataman: (1) Telefol ancestors 'seeded' the landscape with stone adze blades which later became mine sites (Ok Tedi, Porgera, Nena); (2) Kayalikmin ancestors entered Nenataman 'smelling out' the *bangelii* that was the source of the 'gold' at Nena; (3) the Nena mine site belongs to Telefolmin not merely by conquest, but by ancestral right.

A second myth relevant to Nena claims is current in the village of Wabia, where it was told to me by one of the village's leading men. Rather than locating the source of the minerals in the actions of ancestors from Telefolip, it emphasises the linkage between the Nenataman colonists and local spirit beings, a relationship established through affinity and sacrifice. The myth goes as follows:

The myth of Kisimmen

An Ariamuvip man went hunting for wild pig near Ekwai-dubom one night, and after he had bagged one he started for home. As he was walking he noticed a light coming from a cave and came nearer to investigate. He looked inside and saw a woman sitting there making a netbag by the fire, and she spoke to him. 'You already know me – it's me, not some other woman, that you see in your dreams, and I have sent you things. Come sit down with me.' When he dreamed, this man would dream of having intercourse with this woman, and when he awoke he would find bark string she had spun for arrow bindings in his netbag. Then, when he hunted, he found game.

The man was frightened and was worried that she was a ghost or spirit woman. He wanted to run away, but was afraid and so sat down with her. Then he cut some of the wild pig and gave it to her to see if she would eat it raw, but she said, 'Whose is this?' When he said it was hers, she said no, it was theirs and that she was coming with him. So he took her home with him, and when they came to his house he placed the wild pig before the threshold and then they stepped on it as they entered the house. When they got inside, however, a huge thunder and rainstorm came up and shook the house. They went outside again, and he killed a domestic pig and also set it on the threshold. They put their foot on it then, and then went inside. The wind came again and then left.

This woman was Kisimmen, and the wind and thunder were her parents, who had come to take away their pay – the first time they came they were not satisfied, but they were happy after the second pig. Kisimmen could see her parents, but to other people they looked like wind. A Kayalikmin ancestor got this spirit woman, and then we multiplied and became a big population.

The story of Kisimmen came to a Wabia man (also a pastor) in a dream, who then satisfied himself by various tests that the dream was true. Many hold that Kisimmen's spirit parents are still present and active in the Nena area, and that they have the power to protect the interests of their affines and descendants. Others also believe that the Nena project is jeopardised unless it has the support of Kisimmen's parents. This support will be forthcoming if the name of the FMLA and the mining operation are changed to acknowledge them. Hence the desire on the part of some Wabia people to change the name of the FMLA to the Binam-Binip Mine Landowners Association, Binam being the name of Kisimmen's father, and Binip being her mother's name.[31]

Mapping the myths: who's in and who's out

The problem facing Telefolmin in Nenataman is how to situate their claims to mining royalties *vis-à-vis* local contenders such as the Miyanmin, on the one hand, and upland allies in Eliptaman and Ifitaman, on the other. Both myths represent possible solutions by shifting the grounds of discourse from discussions of particular land use histories to overarching claims based upon connections to spirit beings who enjoy a special relation to the land. Thus, through a strategy of mythic encompassment, they pre-empt the possibility of any competing claims based upon prior occupation. They do so, however, in strategically different ways.

The *bangelii* myth synthesises diverse elements of Telefol historical experience in a way that many people find persuasive and satisfying. Capitalising on Telefolip's reputation as the source of ritual knowledge, it is an excellent example of Telefol mythic production at a time when Telefolmin and their neighbours are practising Christians. Quite apart from its implications for mining claims, the myth's integration of Christian and traditional narratives[32] grants it a certain authority which draws upon the local church and the regional prestige of Telefolip. Taking up the traditional theme of the subterranean sources of wealth, its believability derives not only from its affirmation of local hopes, but also from the evidence it adduces. So, for example, the association of valuable minerals with traditionally valuable stone has a tangible form in the shape of the one-kina coin, which echoes the shape of the sacred *tingii* club heads at Telefolip.[33] For local people, the strongest confirmation of this nexus of ideas is based on the precedent of Fubilan (Ok Tedi), which seemingly has the endorsement of the national government (see above). Other confirmations include the fact that the Amosai, an eastern tributary of the Frieda, was the source of *tong* pyrites used for fire-making and is now worked by local people for modest amounts of alluvial gold (see version 1). Finally, some have noted the suggestive presence of a stone axe on HGL's corporate logo, a usage mirrored in the image of a *bangelii* on the logo of the FMLA. In the light of this, the proximity of the Nena deposits to the *bangelii* quarry site seems to speak for itself.

From the perspective of the *bangelii* myth, the benefits from the Nena project should be allocated not to all Telefolmin, but to the Kayalikmin descendants of Wipnagim and Titak-unam in Nenataman, Eliptaman and Ifitaman. This is not simply because of their conquest of Nenataman, which is explained as an *effect* of Kayalikmin rights, rather than their basis. This view thus privileges descent over locality: discussions of land and territory are rendered irrelevant by the prior claim over the minerals themselves.[34] The outcome of this solution would, by incorporating Kayalikmin elsewhere, expand the number of landowners to approximately 2000 or so while excluding Miyanmin and other local claimants.

The Kisimmen myth, by contrast, takes a different tack. It derives its authority not from Telefolip and its mythic traditions, but from a dream encounter between the spirit beings themselves (Kisimmen's parents) and a respected church leader. Local factors clearly predominate, and rather than playing up the reach of Telefolip, it legitimates the claims of its proponents by appealing to a direct link to autochthonous spirit beings in Nenataman itself. Rhetorically, this view serves as an equally strong counter to claims by non-Telefolmin, but would also work to exclude Telefolmin from Ifitaman and Eliptaman.

The resort to mythology draws in part on perceived successes in the Ok Tedi case, and is in keeping with Telefol notions about truth and the source of things. This neatly resolves the problem posed by the recency of Telefol settlement by asserting their claims were in place at the outset. The *bangelii* myth, in particular, draws upon the ritual authority of Telefolip, all the while tacitly acknowledging the military dependence of Nenataman Telefolmin on their upland cousins.

Yet attractive as these myths may be to their proponents, each of them is beset by problems. The *bangelii* myth's emphasis on

Kayalikmin ancestry has an exclusivist thrust, and some have argued that only 'pure'[35] Kayalikmin should qualify, insisting that only those descended in the male line should receive benefits. Others have carried this line of thinking even further by suggesting that a person must have *both* Kayalikmin ancestors *and* ancestresses to qualify as a landowner. There is little warrant for this position in Telefol custom, however, since descent is fully cognatic: individuals carry the *tenum miit* identities of both parents, and there is no discrimination on the basis of patri- or matri-affiliation. More to the point, the pragmatic implications of the purist position are not widely accepted in Nenataman since a sizeable minority of Telefolmin there (particularly in Ok Isai) either have Atemkayakmin ancestry or are descended from assimilated captives.

The myth of Kisimmen meets with its own problems precisely because its central pragmatic assertion would exclude most upland Telefolmin in Eliptaman and Ifitaman. Not surprisingly, many there have dismissed the Kisimmen myth as inauthentic, claiming it has no precedent in local culture and amounts to a story made up for the purpose of cutting them out of any mining agreement. Lacking the authority of Telefolip, it is vulnerable to such challenges and receives little credence outside Wabia.[36]

Configurations of identity in a regional field

Although both the *bangelii* and the Kisimmen myths assert the priority of Telefol rights over their local rivals, they are as much concerned with establishing the terms of relationship between Nenataman and upland Telefolmin as with countering Miyan contentions. In this sense, each proposes an alternative means of defining landowners and generates cleavages in the process. So, for example, the *bangelii* myth's stress on descent from Telefolip denies the claims of locality by its willingness to exclude Telefolmin who are simultaneously descendants of Nenataman's aboriginal inhabitants, and by so doing, opens an embryonic fault-line within the local Telefol community. The Kisimmen myth, on the other hand, flies in the face of the tenet of Telefolip's centrality, challenging Nenataman's historical status as periphery to Ifitaman's core. In thus proposing different ways of distributing mining benefits, these views suggest different configurations of identity for Telefolmin in Nenataman.

What these differing positions expose is an ambivalence located in the tensions between the contrary pulls of place and ethnicity in trying to work out who will get what when mining goes ahead. As we have seen, one tacit model for the distribution of benefits is based on the Telefol experience with the Ok Tedi project, in which mythology served to validate claims based on ethnic identity. From this point of view the *bangelii* myth seems made to order to avoid what Telefolmin call 'the Tabubil sickness' (*sik bilong Tabubil*), namely the awarding of royalties to local claimants (in this case, the Wopkaimin) to the detriment of others with ancestral claims. Pressures for affirming the stake of upland Telefolmin arise not only from myth, however, but from Nenataman's military and ritual dependence on Eliptaman and Ifitaman, and continue today in a number of ways. Eliptaman people, in particular, have been very vocal in their insistence that their claims be recognised, and there have been hints that failure to do so could have serious repercussions ranging from the 'spoiling' of local sago to the razing of villages. Thus broad ethnic claims are situated in a political context, and the power of Eliptaman and Ifitaman is difficult to ignore, since they are a force within the FMLA itself and in the arena of provincial politics. At the same time, organisational forms have emerged which serve to underscore the autonomy of Nenataman Telefolmin *vis-à-vis* their upland congeners. Here the most significant factor has been the local church organisation, the Frieda Baptist Association (FBA). Led by a network of pastors spanning Telefol, Miyan and Paiyamo ethnic lines, the FBA is headed by a Wabia man who has established a small settlement near Frieda strip, directly along the Telefol-Paiyamo frontier. While inconspicuous in formal politics, the FBA constitutes one of Nenataman's most influential and powerful forces and embodies a locally-based focus of identity in which ethnic divisions are downplayed.

Conclusion

A major problem in the Nena case has been the attempt to establish the boundaries of 'landowning' groups where these are inherently ambiguous. Leaving aside the quite substantial matter of Telefol-Miyan divisions, the issue remains stubborn because a number of Telefolmin outside the valley entertain claims which have the effect of blurring the definition of 'landowners', and the incorporation of significant numbers of war captives has had a similar blurring effect on ethnic identities. The problems facing Nenataman Telefolmin in formulating their claims to rights arising from the Nena project, then, consist of two different kinds. They are at pains to assert the priority of their claims when their status as newcomers remains fresh in local rivals' minds while at the same time balancing the claims of upland allies demanding their own share. In a sense, then, it is this indeterminacy that the two contrasting mythological approaches attempt to resolve, but with only limited success. Deciding who will be a landowner thus inevitably becomes an exercise in alignment and self-definition in which the calculus of advantage intersects with questions of identity, all of which is played out against a backdrop of regional political relations. In this context, we should not be surprised if the question of who a landowner in Nenataman is fails to receive a single, unambiguous reply.

While it is easy to see the play of power and interest in this context, it would be a mistake not to recognise that many of the Nena situation's uncertainties arise from the very notion of landownership as the basis for distributing mining benefits. As we have seen, customary land tenure provides an ambiguous guide to this question because a multiplicity of rights along a sliding and somewhat heterogeneous scale is recognised, and this uncertainty is compounded by the fact that none of the traditional usages governing land offers a satisfactory analogue to mineral rights. For this reason it is impossible, even with the best will in the world, to provide a single clear answer to the question of who a landowner is.

Beyond this, however, lie other difficulties embedded in the various strategies the government has adopted in the identification of landowners. In PNG it has often been assumed that the easiest solution to defining landowners is to identify a bounded group, almost always referred to as a 'clan', with a delimited territory. Such assumptions reflect an administrative appropriation of anthropological models of segmentary social organisation[37] which envision a mapping of discrete groupings onto similarly discrete parcels of land. This holds obvious attractions for those seeking a way out of the reality of messy and tangled land disputes, but runs afoul of local custom: Telefol *tenum miit* are not clans with mutually exclusive memberships, and it is normal for people to have overlapping and multiple identities.[38] Another closely related tack consists in attempting to fix ideas of ownership genealogically, implicitly identifying land tenure with inheritance (often with an assumed patrilineality).[39] While this meshes with some aspects of the local situation, it neglects the force of claims founded less upon inheritance than upon histories of work or military assistance, thus leaving players out of the picture who are nonetheless powerful in regional politics. At the same time, local perceptions of a patrilineal bias can give rise (as in case of the partisans of 'pure' Kayalikmin descent) to new kinds of exclusionary claim based on a reification of privileged genealogical ties. Although lacking the authority of established custom, such moves are perhaps to be expected in a context where getting a share of the benefits depends on coming up with the 'right' answer to outsiders' questions about who one is. The notion that land is vested in clans does less to resolve questions of land ownership than to inject an additional complication into what is already a murky situation (see Zimmer-Tamakoshi 1997, Guddemi 1997).

If there is a practical lesson to be learned from this case, it is that a unitary concept of landownership poses a serious obstacle to arriving at a workable settlement of claims in Nenataman. To say this, however, is not to say that some form of settlement is beyond reach, providing there is a willingness to come to terms with a sliding scale of claims and entitlements. In fact, the multiplication of varieties of benefits and compensation characteristic of contemporary mining agreements, ranging from royalties to the provision of health and educational services, holds out just such a possibility. It seems clear that, in Nenataman at any rate, any settlement will have to be prepared to countenance a diversity of claims with a corresponding range of different benefits if it is to have any hope of survival.

Postscript

After all the above sections of this chapter had been written (May 1996), further developments on the Nena scene came to light indicating that the reformulation of identities underway in 1995 had taken a new and interesting turn. In the wake of unsuccessful attempts to arrive at an agreement between Miyan and Telefol claimants to mine sites in Nenataman, a coalition of people claiming descent from the aboriginal inhabitants declared their intention of seeking registration as landowners as members of the 'Untoumin clan' (Gardner and Togolo, pers. comm.). While details are sketchy, the information available suggests that this group includes descendants of war captives among both Telefolmin and Miyanmin, along with the Paiyamo, who claim to be the nearest living relatives of the Untou.[40]

This development now offers the prospect of a multi-ethnic 'clan' whose members speak three languages from two different language families cross-cutting four different villages, thus scrambling the prevailing pattern of cleavages and alignments in Nenataman. As such, it represents a culmination of tendencies to assert the priority of territorial affinities over larger categories of ethnic identity. Here a network of cognation and affinity across ethnic lines plays a role, and since the end of warfare incorporated captives have served as links between different communities, a pattern strongly reflected in the role such people play as pastors in the FBA (see above) and in official leadership positions.[41]

What is striking about the evidential basis for the existence of an Untou clan is that, like the arguments examined elsewhere in this chapter, it is couched in terms of identity rather than histories of land use. This approach does so, however, by reifying genealogy in a way that ignores adoptive kin relations. This departs from local custom – since adoptive kin count as real kin among Telefolmin, Miyanmin and Paiyamo (cf. Sagir, this vol.) – and very likely reflects local perceptions of the importance the government attaches to genealogical proofs in statements of claim.

For Telefolmin, the implications seem to be twofold. The first is the rejection of the claims of Eliptaman and Ifitaman Telefolmin outright, and the second is a further division among Nenataman Telefolmin, particularly within Ok Isai. Here it is tempting to view the re-emergence of the Untou among Telefolmin as a reaction to the arguments of the *bangelii* myth, turning the tables on the partisans of 'pure' (that is, patrilineal) Kayalikmin descent by asserting the priority of 'aboriginal' claims.[42]

The effect of all this is as yet unclear, though it is doubtful that it will contribute to a resolution of the problems of reaching a workable mining agreement, if for no other reason than that the exclusion of a significant number of Nenataman residents dooms any agreement to failure on purely political grounds. These developments may be regarded, however, as further consequences of an approach to land issues that phrases matters in terms of property rather than an accommodation of the interests of local groups within a field of political relations. Insofar as this is the case, the resurrection of the Untou represents yet another attempt to recast identities as local people try to reshape themselves to fit the definition of landowner as they perceive it. While unlikely to succeed for a number of reasons, this move is sure to have an impact on local politics for some time to come.

Acknowledgments

This chapter is based upon the results of six weeks' fieldwork in Nenataman conducted in mid-1995, and draws upon earlier research conducted among Telefolmin since the mid-1970s. My work in Nenataman and adjacent areas was carried out in the course of a land claims consultancy (at the invitation of the Frieda Mine Landowners' Association) for Highlands Gold, Ltd. I worked as part of an anthropological team which also included Don Gardner, George Morren, and Rune Paulsen. Our tasks were organised along ethnic lines, with the following areas of responsibility:

Gardner:	Miyanmin, Sepik Iwam, Owininga and overall coordination
Jorgensen:	Telefolmin, Paiyamo
Morren:	Miyanmin
Paulsen:	Tunap Iwam

I am grateful to the other members of the team for their insights and to Imke Swart Jorgensen, John Gehman, Paula Brown Glick, Don Gardner, George Morren, Laura Zimmer-Tamakoshi, Anton Ploeg, Phillip Guddemi, Barry Craig, Stuart Kirsch, Colin Filer, Joel Robbins, Andrew Strathern and an anonymous reviewer for comments on previous drafts. Tom Moylan generously made his earlier unpublished work in Nenataman available to me. Many Highlands Gold staff were helpful, including Mel Togolo, Tony Friend, Arnold Smare and Paul Green. Danny Lane, John Yapi, Bob Onengim and the late Levi Binengim, all of the FMLA executive, were invaluable in helping to understand their view of the local situation. Of the local people who assisted me I would particularly like to thank Dusamnak, Naron, Peter Mansimnok, Muie, Megilkep, Tigap, Wesani Iwoksim, Suumengim, Fobayok, Yemis, Bumanok, Bisomsep and Beksep.

This chapter is a slightly revised version of one that appeared in *Anthropological Forum* in 1997. The editors of the volume and I thank the editor of that journal for permitting us to republish it.

Notes

1. These issues intersect with broader questions concerning the status of contemporary demands for 'compensation', in which efforts to assert various kinds of material interests are cloaked in a mantle of 'tradition' which may nonetheless be far from customary (Strathern 1993). For an account of conflicting claims and interests in the context of conservation planning, see Kirsch (1997a). Recent overviews of the mining and petroleum industries in PNG are available in Cook (1996a, 1996b).
2. This has changed with time, and Telefolmin – still the largest component of the labour force – now occupy many highly skilled positions at the mine.
3. Language and culture here, as elsewhere in PNG, are not always isomorphic. See Jorgensen (1996) for details.
4. Morren (1986b) has argued that Miyanmin resistance to the extension of colonial control until the 1960s was a continuation of resistance to Telefol expansion begun a century earlier.
5. I should add that Telefolmin did not question the Wopkaimin right to recognition, but rather complained that this was unduly restrictive, since Fubilan belonged – in their view – to all of Afek's children, including Telefolmin.
6. Vail (1995) writes of a similar movement among the Huli in relation to the Mount Kare mine.
7. Such status had in fact already been recognised in part of the original agreement providing for preferential hiring for people of the Telefomin District of Sandaun (West Sepik) Province.
8. Christianity was originally introduced to Telefolmin by the Australian Baptist Missionary Society in the 1950s. Since the late 1970s, most Telefolmin have been professed Christians practising their own version of Christianity, formulated as *Rebaibal*. For details, see Jorgensen (1981*a*).
9. I would estimate that between one-third and one-half of Nenataman Telefolmin can trace descent from war captives. It is arguable that the assimilation of captives was necessary for the demographic viability of Telefol settlements in Nenataman.
10. Claims are divided among siblings equally, with no bias in favour of seniority or sex.
11. Nenataman lies between 100 and 800 metres above sea level; Eliptaman and Ifitaman range from roughly 1100 to about 1900 metres above sea level.
12. Sago was also a necessary ingredient in junior male initiations.

13. Given the precedent of violent action surrounding the Bougainville and Mount Kare mines, this stance has obvious implications for the future success of mining ventures in the area, a point that some individuals made explicitly with reference to Nena.
14. A notable exception here was ritual, for Nenataman settlers still sent youths to Telefolip to be initiated.
15. There is no articulate body of custom pertaining to such situations, but there are other examples of similar 'sponsorship' by original inhabitants, both in Nenataman and in the earlier Telefol conquest of Eliptaman. One concern was to learn about taboo places (*amemtem*), names of local spirits, location of 'bad water' and other features of an unfamiliar landscape, although this clearly played no role in Obulum's case.
16. In earlier (pre-HGL) days of mineral exploration at Frieda Base, Telefol employees often worked without lunch breaks (Don Gardner, pers. comm.). Company officials justified this by telling people that they were working to develop 'their own mine', a view local people took very much to heart. The shift of focus to Nena Mountain, which raised new questions of land claims, was disconcerting, to say the least.
17. The ironies and contradictions in this situation should be self-evident. From the perspective of Highlands Gold and of the Frieda Mine Landowners' Association, our brief was to assemble documentation of land use and land rights, and the kind of evidence envisaged was for the most part understood as genealogical. We were also charged with communicating the concerns of local people in our consultants' submission, touching upon matters involving employment, training, provision of services, worries about a possible influx of outsiders, and so on. At a more general level, HGL clearly hoped that our work would reveal a basis for arriving at an accommodation of mutually conflicting claims over the proposed mine site in order to avoid strife with local people should the project go ahead, an aim the FMLA appeared to share. But in my conversations with individuals in a number of different villages it was evident that their first priority was simply to assure that their own claims would be taken seriously. In a number of cases this was coupled with assertions meant to disqualify the claims of others with whom I had already spoken. Under such contentious conditions, the role of advocate was problematised by the number of different groups of people wanting me to act on their behalf, and I found myself continually obliged to issue disclaimers about my ability to determine the course of events and stressed that my role was to assemble and transmit villagers' accounts. Since most of the parties to these discussions were well aware of the local

configuration of interests and forces, an appeal to a pragmatic spirit of compromise was greeted with a measured realism. Nonetheless, it was evident that not all hopes would be realised: my lot in this situation was not a happy one, and the strains on the role of the 'honest broker' were severe. It seemed to me, however, to be the best one could hope for, and better than any of the available alternatives.

18. This is not by any stretch of the imagination the whole story, and Miyanmin have, needless to say, their own version of events here.

19. This is a calque upon vernacular distinctions between stories of the past (*sogaamiyok sang*) and sacred myths (variously *weng amem* and *Afek sang*). See Jorgensen (1990b) for details.

20. For other examples of mythologies surrounding resource development in PNG, see Biersack, ed. (1995), Clark (1993), Jorgensen (1990b) and Weiner (1994).

21. The reference here is to a netbag of sacred relics (*men amem*), including Umoim's bones. It is in the possession of the people of Telefolip.

22. Telefol villages in Eliptaman and Ifitaman.

23. Here the narrator used one of her secret names as a way of validating what he said.

24. This is an apparent reference to a sacred tree at Telefolip in which Afek made Umoim's mortuary platform after she killed him.

25. This sequence is patterned on the details of the traditional account of Umoim's death.

26. Shells are held to originate from maggots in Umoim's rotting flesh, underscoring the associations of wealth with death and regeneration. They entered the area along trade routes from Wopkaimin, passing by Fubilan.

27. Alternately, a stranger from Oksapmin (in the east, where the sun rises) is the victim. This bears an intriguing resemblance to accounts of the killing of a human victim during Huli and Duna rituals. See Glasse (1995), Strathern (1995).

28. The linkage with the two women comes full circle with the later deposit of adze blades at their respective places.

29. In the traditional rendering of the tale of Kwiinagim, his bones become powerful relics at Telefolip. The same fate awaits Umoim, but he also creates the source of *fubi* blades and shells.

30. Telefol, *miit* or *magam*; Pidgin equivalents here are *mak* and *as*.

31. In fact, these are only cover names, the true names being too secret to reveal – but it is felt that they would recognise this gesture and smile on the Nena project.

32. Here knowledge of the Bible and use of esoteric names for characters in the myth plays an important authenticating role.
33. It is probably worth pointing out that the Telefol word for money is *tumoon*, 'stone'.
34. There is here an uncanny parallel to legal views which formally allocate subterranean mineral rights to the state, regardless of the ownership of the surface.
35. The English word 'pure' was used.
36. It also finds favour in Ankem (Ifitaman), where some of Kisimmen's descendants currently reside, and rumour has it that Ankem people were consulted in the final formulation of Kisimmen's story.
37. It is tempting to regard this, at least in part, as a residual effect of the Australian practice of providing anthropological instruction at the former Australian School of Pacific Administration (ASOPA). Here, as elsewhere, anthropological models may take on a life of their own when deployed in legal and political contexts far removed from their original sources. See Biersack (1995*a*), Allen (1995), and Vail (1995) for an extended critique of the segmentary model among the Huli and Ipili, who are the major landowners in the vicinity of the Mount Kare mine.
38. The same objection applies to Miyan descent categories (*miit*) and Paiyamo categories (*pesinalage*) which might best be termed 'house groups'.
39. This is reflected in the original definition of our consultants' brief (see above).
40. There is some justification for these claims on linguistic and ethno-historical grounds.
41. Indeed, strikingly so: the intermediary role of former captives or their descendants has flourished, accounting for the *luluai* (government headman) and *tultul* (his deputy) in Ok Isai, as well as the leading candidate for leadership in the recently introduced community government system. Clearly, such status has not been a handicap.
42. The Untou case bears an intriguing family resemblance to the implications of the Kisimmen myth, a resemblance that is strengthened by a genealogical coincidence: a comparison of genealogies from different locations reveals that the woman identified as Kisimmen is known elsewhere as Finamkonip, who was taken captive in a raid against the Untou in the previous century.

5

CONTINUITY AND IDENTITY: MINERAL DEVELOPMENT, LAND TENURE AND 'OWNERSHIP' AMONG THE NORTHERN MOUNTAIN OK

Don Gardner

Disputes over the allocation and distribution of benefits are, of course, an apparently inescapable dimension of resource development in Papua New Guinea. A growing body of literature, produced by experienced and distinguished researchers, has addressed this issue, either in particular cases or from a more general perspective (for example, the papers in Connell and Howitt 1991; Filer 1990; Jackson 1982, 1991; Welsh 1987). This chapter examines a particular dispute over the identity of 'landowners', although my aim is to suggest a broader framework for the interpretation of such disputes, and to subvert constructions of the motivations of disputants that are apt to play a central role in the negotiations for their resolution.[1]

The Nena and associated gold and copper deposits are located in an area that, all contemporary residents agree, some generations earlier belonged to a group named the Ontouten (see Map 5). Twenty-five years ago, after lengthy investigations and negotiations between Miyanten and Telefolmin peoples currently living in the area, the developer and the PNG government identified the Miyanten as the owners of the land on which the deposit is located. More recently, the Telefolmin, apparently reneging on the earlier agreement, claim to be the true 'owners' of the land around the prospect. In this chapter I outline the land dispute between these two Mountain Ok peoples and the social organisational and epistemic contexts in which it was conducted. Specifically, I would like to try to make some sense of this dispute in terms that go beyond the attributions of motives of avarice that struck most of those involved in the dispute as the most compelling explanation of what occurred.[2] I will suggest that the basis for a more satisfactory explanation of the impasse that developed is provided by two sets of considerations: first, the peculiarities of the organisation of communities and subsistence

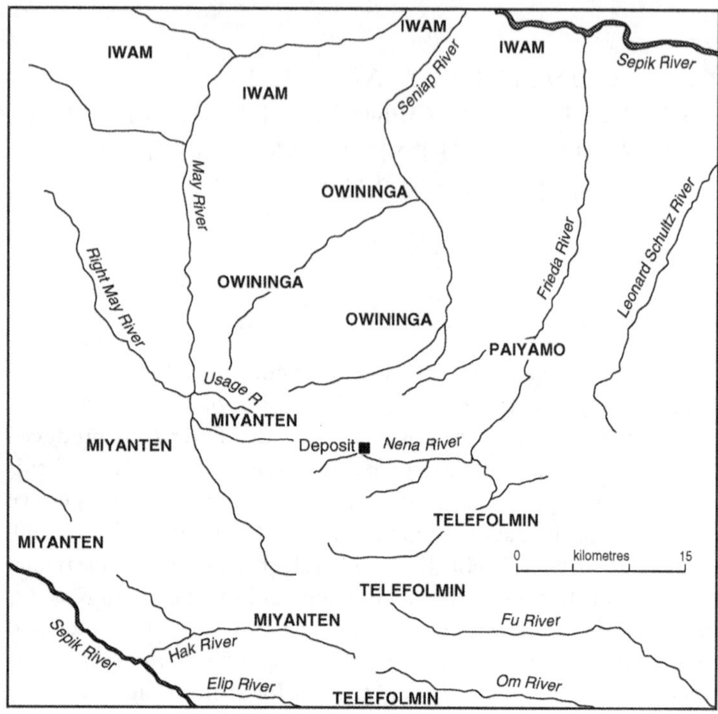

Map 5: Sketch map of the area and peoples around the Nena deposit.

production, which led local groups to pursue inclusive rather than exclusive recruitment strategies; and second, the cosmological context of ideas about knowledge and truth. In the process, I wish to show how discourse about mining and comparable resource developments, which is invariably framed in the language of 'rights,' tends to pre-empt the ethnography, to naturalise Western legal perspectives and to minimise the efforts we need to make to achieve our expressed aim of accommodating the systems of other peoples. I shall suggest that social relations of land use among the Mountain Ok, and the way knowledge-claims are construed and evaluated, together with the state's demand to understand pre-contact practices in terms that can be accommodated within a discourse of 'rights', produce questions whose answers are radically underdetermined by any possible historical precept or practice, so that all options are open. Any resolution must be contingent with respect to all endogenous beliefs and practices, saving that which constitutes the profile of world events as a consequence of ancestral-agentive power.

I should make it clear that the scare-quotes in the title are not meant to indicate that I see any possibility of doubting that the land at issue belongs to Mountain Ok people, or that within our legal and ethical discourse they have rights which we must take into account. Coming from a land whose indigenous people were until recently denied land rights on the basis of a doctrine of *terra nullius,* I have no wish to provide any sort of ammunition for those who would challenge the claim that these lands belong to their indigenous occupants in the strongest possible sense. My aim is simply to suggest that by attempting to strait-jacket social systems like those of the Mountain Ok within a framework of specific legal rights, we prevent ourselves from appreciating the complex processes which distribute people in space: as a consequence, we often craft legal settlements and solutions that do not work.

The area and the dispute

Political geography of the region

The Telefolmin, the largest and most central of the Mountain Ok-speaking peoples, have been expanding the territory they control for the past 300 years (Hyndman and Morren 1990:21-2). Over that period, they have also come to be regarded as the ritually pre-eminent group in the region: the power of their ritual practices, centred on the cult house built by the founding ancestress herself, is acknowledged as superior by many other peoples of the region, including some of their bitterest enemies. The bulk of the Telefol population is still found in two large central valleys: the Ifitaman, which is the original centre of expansion, and the Eliptaman, which was annexed in the second half of the 19th century. The Nenataman Telefols, who occupy areas in the Frieda and Nena valleys (to the north west of Eliptaman) that came under their control late last century, are the most northerly and the most remote of these expansionary segments. The most consistent opponents of the Telefol push northwards have been another Mountain Ok group, the Miyanten (Morren 1986a).[3] The Wamei, the easternmost Miyan group, and their allies fought continuously to keep the Telefols from encroaching on the territory they controlled. Today, almost four decades after the arrival of colonial forces, which inaugurated a period of cooperation between the two groups, the Nenataman Telefols and the eastern Miyanten find themselves contending for recognition as 'landowners' by the PNG state and the company developing the Nena deposit.[4] As everyone realises, being so recognised is of enormous material consequence, given current PNG practice concerning the distribution of compensation and royalty payments.

Mineral exploration and the development of a new future

The Nena Prospect is located in the valley of one of the rivers draining Mount Stolle, which dominates the fringes of the central ranges north-west of Telefomin (see Map 5). In addition to the Telefolmin and Miyanten, the Nena impact area embraces several other groups: Owininga (Left May) peoples occupy lands to the north, and to the west and north west are Paiyamo (Leonard Schultze) people, while the large rivers to the north east of Nena (and on the Sepik itself) are occupied by various Iwam sub-groups. Although these other cultural groups are vital to the overall picture at Nena, I shall not have much to say about them.[5]

A total of about 600 Telefols, occupying two villages, live in the Nena area. Miyan claimants – all of whom identify as Wamei – numbering a total of some 380, are rather more dispersed; one group occupies the headwaters of the Nena River itself, above the deposit, whence they returned about 10 years ago, while two other Wamei segments (one of which is now known as Soga) occupy settlements in other valleys up to three days walk from the deposit.

In many respects the contemporary scene in the Nena impact area is incomprehensible without some appreciation of the history of exploration. After initial investigations by administration geologists had established the potential of the April and Frieda systems, in the 1960s, Prospecting Authority 58 was acquired by the Carpenteria Exploration Company, which began its exploration program in 1969. Initial exploration led the company to concentrate its efforts in the Frieda system, where the Nena and its tributaries seemed particularly promising. Later, Frieda Copper (as it had become when CEC was joined by Japanese and German companies) built the Base Camp on Prospect Creek, a southern tributary of the Nena, still used by Highlands Gold. At the time of construction, the settlements nearest to Frieda Base Camp were those of the Telefols of Unamo and Wabia, in the upper reaches of the Frieda. Men from these villages, and their relatives in the populous Elip valley, quickly came to dominate CEC's workforce. The picture of land ownership as it presented itself to CEC seemed clear: the land around the airstrip and to the north was owned by the Paiyamo of Bapi, while the lands closest to the Base Camp were owned by the Telefols of Unamo and Wabia.

Exploration in several areas around Frieda Base yielded results that produced a spirit of optimism about the prospects of there being a significant copper mine at Frieda. The early optimism of Frieda Copper, which is evident in company memos and the patrol reports of the 1970s, certainly communicated itself to the people in the eastern portion of the impact area, especially the local Telefols, and became part of their own vision of their future.

Through the seventies and early eighties other prospects in the vicinity were investigated, including some on the northern side of the Nena Valley, one of which was the ('Nena') deposit now the subject of the developer's close attention. Land ownership was only really an issue in relation to these northern prospects and those in the uppermost tributaries of the Nena system. Without detailed historical information, and in the absence of any other groups, the claims of the Telefols to the land around Frieda Base were taken at face-value, as were the claims of the Miyans to the northern and upper sections of the valley that – at the time – were peripheral to the main exploratory efforts. Although the exact boundary was unclear, a settlement was reached after two government patrols and discussions with both sides in the late 1970s. These negotiations, and a subsequent genealogical study by Tom Moylan, confirmed that Miyans and Telefols were ready to cooperate with one another. The Miyans did not dispute the Telefol claim that they had driven out or killed most of the Ontouten, nor that they had incorporated several significant captives into their own communities. The Telefols did not, for their part, contest the Miyan claim that the Ontouten was a Miyan group with close kin links to other local Miyan communities (notably those of the Wamei). Both sides acknowledged that the exploration area could only be used for hunting and other short term activities because of the danger of attack from enemy raiders. Both were ready to agree that since the entire area had been a buffer zone following the demise of the Ontouten, some sort of sharing of the occupation fees was appropriate: the Telefols received fees from eight and the Miyans from one of the areas being drilled.

Subsequent events have transformed this picture, but I want to stress here the horizon of expectations that exploration has set up and that underlies the current land dispute between the Telefols and Miyans: it is a key fact that the current generation of young adults has grown up with the expectation that there will be a mine in the area. The disappointment felt by Telefols, Miyanten, Paiyamo and Sepik Iwam when Frieda Copper withdrew in the 1980s must have been enormous, but that was expunged by the arrival of HGL; once more expectations ran high, even if they were tinged with apprehension about whether a mine would eventuate. The expectations of local people, in addition to their labour inputs and cooperation they feel they have shown, have given them a very strong conviction that they too have made a tremendous investment in the future of the development. These expectations were evident in the responses of the people of the Frieda when, in early 1995, the developer mooted the possibility of moving the bulk of the infrastructure from the Frieda to the May River; people responded with anger and disbelief, likening their situation to a man who has raised a daughter only to learn that he would have no share of her bridewealth.

It is against this background that the current position of Telefols is to be interpreted. The Telefols now maintain that the Nena deposit is on land that they annexed in their rout of the Ontouten; that the Ontouten were not a Miyan group at all, because they spoke a different language; and that aside from the women and children they captured and incorporated into their own communities no Ontou survived. By 1995, when Dan Jorgensen, the most experienced of the Telefol ethnographers, arrived to carry out some research, Nenataman Telefols had travelled to Eliptaman and to the ritual centre of the region, the Telefol village of Telefolip, and amassed secret sacred texts suggesting that the Telefol claim could be justified by reference to cosmogonic events in addition to the more recent history of annexation.[6]

The Wamei Miyans, by their own and other Miyan accounts, have occupied the upper May and Nena valleys since the time of the founding ancestors, except for periods when strategic withdrawals were necessary. They claim to have been closely related to Ontouten inhabitants of the lower and middle Nena, and to have absorbed the remnants that fled from Telefol raiders. Indeed, they claim that Ontouten was but a branch of the Wamei, descended from the same son of the founding female ancestor. They do not dispute their opponents' claim that the Ontou did not speak the Miyan language, but still maintain that they *were* Miyan (although, it must be added, this would not have been the way the identity was framed since this name is a recent introduction). Indeed, they claim that the language spoken by the contemporary Miyan is relatively new, and differs from that given to them by the founding ancestress, having been adopted from the people of the headwaters of the River Fu, with whom they enjoyed friendly and frequent contacts before the northwards expansion of the Telefols drove them apart.[7] Other historical accounts show that speaking the same language was neither necessary nor sufficient reason for good relations between groups, and, conversely, that speaking a different language never prevented groups from forming important alliances. The Wamei Miyans retain ancient pestles and bones, dating, they assert, from the time of their founding ancestor, which establishes their title to the Nena and other valleys. In the course of his research for HGL, George Morren collected long, detailed accounts of Wamei history in the area, but it is hard to reconcile his and Jorgensen's accounts of even quite recent periods.

It is fair to say, I think, that most of those connected with Nena, with the exception of those who are Telefol, tend to side with the Miyanten. The reason for this is that the Telefols are seen to have 'changed their tune' about the ownership of the deposit, and most people feel they can provide a ready explanation for this: their actions are interpreted as the triumph of cupidity over virtue. I confess that I was initially inclined to a version of the same interpretation. But I spent several days living and speaking with Telefols about their claim, and in relaying to them my findings in the archives of the Mining Warden's Court, which showed that aside from the initial agreement, they had affirmed (by deed if not word) on several different occasions between 1979 and 1984 the Miyan version of things, which had the Telefols receiving occupation fees for the prospects south of the Nena River and the Miyans for those north of it. I was struck by my interlocutors' genuine dismay and the seriousness – without any trace of behaviour that might suggest they were 'trying it on' – with which they considered the news and tried to come up with explanations for why the Mining Warden had got it wrong. Some maintained that they had been saying what they are saying now (in much less detail) since the question had first been broached, others that the officers who had carried out the investigations had asked the wrong senior men, a view given credence by the wealth of material since collected from elders at Telefomin and Eliptaman.

The anthropological question I pose here concerns the orientation to the world, groups and knowledge that might account for the Telefols' present conviction. Of course, at one level, this question has an obvious answer: what we are faced with is the operation of mythic consciousness that anthropologists have probed for so long. Indeed, Jorgensen has presented careful analyses of the operation of Telefol mythic consciousness in various contemporary contexts (1990a, 1990b, 1996, this vol.). My own purpose is somewhat tangential to this enterprise. Anthropologists – quite rightly – tend to bracket questions of motivation when dealing with the bricoleur, whereas I try to place it in the foreground, since this is precisely where it appears when government and company officials are faced with expressions of local cosmological views. I do not offer a psychological account, in the sense of generalising across individual motivational states, but merely try to sketch a context (derived mostly from my understandings of Miyan) which I hope diminishes the appeal of the usual explanation for the current Telefol position.

Miit as concept and in practice

As most Melanesianists now know, the Mountain Ok peoples often constitute themselves as 'Min,' a designation which refers not to the obvious linguistic and cultural affinities between them – at least not initially – but to their being the descendants of a single female, or a group of sisters. The pervasiveness of maternality as a principle of generation of all sorts of important natural and non-natural practices, objects and living creatures is remarked by (almost) all ethnographers of the area. It would not be going too far, I submit, to suggest that for the Mountain Ok 'phylogeny recapitulates ontogeny;' the processes of the world – in itself, and as lived by individual humans – always take place in the context of a mother.

But if the ultimate origin (*miit*: 'kind,' 'root' or 'base') of all Mountain Ok peoples is a woman, the immediate source of identity for living individuals is the named 'man-*miit*' (*tanum miit* in Telefol; *naka miit* in Miyan). It is at this level that the dispute between the Telefolmin and Miyanten is couched, since this term is basic to discourse about territorial communities (usually called parishes in the literature). For example, the western Miyanten population (the one

I know best) is divided into sixteen autonomous territorial units, each of which bears a *miit* name: the Ivikten, Fiyariten, Futibinten, and so on. If we note that the Miyanten had no inclusive term for the population as a whole prior to the arrival of colonial powers, it can appear that the Miyanten, like the Atbalmin, 'are not one people, but many' (Bercovitch 1989:4). These facts, together with the Miyan preference for local endogamy, make sense of the Australian administration's former practice of designating such units 'sub-tribes.'

The concept of *naka (tanum) miit* is explained by reference to ideas about the genesis of bodily constituents, which we will consider briefly. In explicit formulations, people say that a foetus is initially formed by the interaction of male and female genital secretions. When the foetus is formed it is infused with the mother's menstrual blood.[8] The significance of these ideas is that children are constituted from their father's bone and their mother's blood and that these substances give corporeal expression to lineal continuity; at least, in principle, they can. In reality, in so far as continuity is represented somatically in general contexts, it is almost exclusively in terms of bone, the male contribution to bodily substance. Blood, and connections established by it, are usually discussed in the context of relations among *miit*, whose identity may be rhetorically defined by the transmission of male substance, bone (but see below).

As noted already, each Northern Ok territorial group bears a *miit* name, usually that of the most important *miit* (which is not invariably the best-represented). It is usual, however, for such groups to comprise individuals of several different 'kinds.' In addition, many of these territorial groups are internally divided into residential groups led by a pre-eminent man or *kamok*. The desire of these men to augment their followings goes some way to explaining the *miit* heterogeneity of territorial communities that are publicly designated by a name that would indicate homogeneity. However, given the explicit preference for a strategy of local endogamy (which in my western Miyan samples runs at slightly over 50 per cent of all marriages, but is greater in areas where local territorial populations are larger), most people are affiliated with their territorial *miit* through each of their parents. When exogamous marriages take place the offspring are likewise affiliated to both *miit*. In my Miyanten samples, some 25 per cent of men reside non-patrilocally, a fact which reflects the openness of local groups to new members, as does the frequent occurrence of dual residence, serial residential shifts and other facets of northern Mountain Ok social organisation.

This fluidity needs to be interpreted in relation to subsistence production. Although northern Mountain Ok population densities vary from group to group, and are much higher in the central valleys of the Telefols than on the fringes (see Hyndman and Morren 1990: 16-17), all of them pursue a low-density, low-intensity, land-extensive hunter-horticulturalist mode of subsistence that contrasts with subsistence systems elsewhere in the New Guinea highlands.[9] Hunting and horticultural techniques, as well as the preference for local endogamy and the need for military security, mean that small groups will always face more difficulties than larger ones. Accordingly, local communities are open to the in-migration of families and even whole segments of other parishes.[10] Almost any kind of connection with a group, or one of its members, will suffice as a basis for admission to a parish. Past movements of individuals and groups have resulted in considerable heterogeneity of parishes that are routinely designated by a single *miit* name. So, while a community's members may be identified with a single, 'public' *miit* name for the purposes of discourse about whole groups (especially in the discourse of socially remote communities), genealogical investigations show that individuals are liable to identify with a whole range of 'kinds.' An individual from a remote group who is captured or voluntarily joins a community, will be said to have 'come inside' the *miit*.[11]

The processes of social life, then, produce genealogically complex and heterogeneous communities. Given the preference for local endogamy, almost everyone descended from a neolocal member of a given community will be able to point to a male or female ancestor who is of the dominant *miit*, the name of which the parish bears. Hence, most ethnographers characterise these societies as cognatic. Yet, this does little to indicate the complexity of social process in the region. For example, in the Mountain Ok peripheries, on the edge of mountain fringes, groups and families have very high levels of residential dispersal and mobility, and whole communities, or parts thereof, shift frequently. (For example, in a 1986 survey, I found that 118 individuals over the age of 30 years named a total of 89 different settlements as their places of birth. Even allowing for the deficiencies of such a survey, this indicates a very high level of dispersal and mobility.) Under these conditions, and given the competition among leaders for personnel, the *miit* identities of individuals and communities may become highly problematic and subject to contestation, as segments of *miit* come together and others split to reside in different, often distant, territories. Ultimately, the identity of particular communities can be set out in the narrative history of its past and present members, their conquests, losses, movements, and so on, but these identities are, in the end, always designated using *miit* names, with their lineal, essentialising connotations. All this indicates that the Mountain Ok *miit*, like many of the entities that constitute human worlds, including persons, clans, societies and nation-states, are essentially historical continuants, which means, among other things, that their identity can be captured by no set of conditions, no matter how exhaustive, that excludes the appropriate causal temporal sequence. *Miit* names refer to the products of intricate historical processes that, through the particularities of individuals bearing specific genealogical identities, also implicate the concept of *miit*. Accordingly, the *miit* figures in Mountain Ok social life in two analytically distinguishable ways (as comparable constructs do elsewhere in New Guinea).

Morren, writing of the eastern Miyanten, has suggested that descent is not important in understanding the *miit* (1986:175) and he prefers the term 'parish', which has definite territorial connotations, in speaking of local groups. Rather than a descent structure:

> It is a matter ... of residence and recruitment with individual parishes, as corporate entities, acting to conserve and expand their populations through endogamy, reciprocal interparish marriages, invitations for outsiders to contract uxorilocal marriages, or the taking of captives in war. [Morren 1986:175]

In a similar vein, Ruth Craig, in discussing Telefols of the Ifitaman Valley, writes of the *miit* name as having either a descent or a local territorial connotation, depending on context (1969:177). Because the territorial usage is the most frequent, she too refers to *miit* as parishes. Jorgensen, in his writings on Telefol society, has stated that 'there are no descent groups' (1988:261), a claim amplified thus:

> Although there exist named cognatic descent categories, these do not govern marriage, have no corporate characteristics and do not form a matrix for collective life. [Jorgensen 1991b:261]

While there is clearly much to commend these views, I would argue that the notion of *miit*, and the lineal implications of the conception theory, do play a crucial if complex role in 'the matrix of collective life,' for they are significant aspects of the way Mountain Ok communities represent and enact their practices. Interpersonal and inter-parish relations would have entirely different profiles in the absence of the practical and rhetorical resources these notions provide. Indeed, in general discussions about *miit* and their membership, people will often suggest a paradigm that looks patrilineal: the *miit* is frequently defined as the sharing of bone (thus, *on miit*, 'bone *miit*,' can be used as a synonym for *naka miit* among western Miyanten). Bone is a conventional by-word for strength, and its continuity and strength figures prominently in the region's sacred cult, with its emphasis on origins and primordial powers. Men often make disparaging contrasts between the side of the father – of bone – and that of the mother. Even a sister's son retrieves and cherishes the bones of his dead mother's brother. A ribald analogy men sometimes draw to underline the similarities and differences between the 'right hand side' of men and the 'left hand side' of women, likens them to the difference between a penis and a clitoris. Miyanten, at least, posit a structural and morphological identity between these two organs (both are emphatically stated to have a bone) so that the clitoris is thought of as a diminutive version of the penis.[12] Agnates sometimes jocularly refer to one another's mother's brother as 'your clitoris'.

My colleagues are surely correct, though, in stressing the lack of any principled or necessary connection between the distribution of personnel across social and geographical space and the concept of *miit*. The conception theory and the notion of *miit* it rationalises, it must be conceded, have only a partial role in articulating relatedness: their potential for systematic and comprehensive discriminations in the field of kin is unrealised in a social context where residential openness and commitments to co-residents impose their own imperatives. Nevertheless, the concept of *miit* and the doctrine of continuity through bone and blood play a significant role in the identity of local territorial communities and in their interrelations, as well as in notions about the somatic constitution of individuals. It is tempting, under these circumstances, to suggest a parallel between *miit* identity and national identity as characterised by Renan, viz, that it depends upon amnesia, but that would not be entirely apt; for, when it comes to community recruitment and its internal composition, it is not so much amnesia as indifference to the provenance of their members that characterises Mountain Ok attitudes to political communities. While communities gain and lose members as the ebb and flow of micro-political processes induce members to make temporary and permanent residences in other places, any given group is faced with the imperative to consolidate its own membership and make connections with individuals elsewhere in an effort to increase its strength.

Other ambiguities beset the question of exclusive group membership and ownership as it is presented in the context of mineral development. For example, the descendants of the Telefols who now occupy the Nena area were only able to clear out the Ontouten and keep the Miyanten at bay because of the military aid given to them by large numbers of their Eliptaman and Ifitaman relatives and allies. Because of the distances involved these allies spent considerable periods in the Nena area. Some of them planted tree crops and bamboo there, and their descendants still come to attend these, or simply to hunt, clear a garden and stay a while (Jorgensen, this vol.). Some Eliptaman Telefols have always spent considerable periods – months at a time – in the Nena area, and some of them have a house or hearth there. Similar instances of this sort of thing could be cited on the Miyanten side too. Dual residence is a characteristic feature of northern Mountain Ok social life. So, for example, in 1986, of 123 western Miyanten nuclear families with both husband and wife still living, 61 had a secondary residence in another settlement. Aside from dual residence proper, and merging into it, extended visits (sometimes many months in duration) to other settlements are also common. Under these circumstances, an individual's identity is a complex matter.

The simplicity of the notion of *miit* entailed by the conception theory, then, belies the complexity of its overall place in the processes of social life. This is not unusual, of course, especially in the New Guinea, where notions like 'loose structures,' the gap between descent as principle and descent as idiom, and the priority of patrifiliation over patriliny were developed precisely to account for the perceived juxtaposition between the way social life is lived and the way it is represented. In the Mountain Ok area, the preference for endogamy means that a crucial context for the fixing of group identity found elsewhere in New Guinea is lacking, and the effects of this are compounded by the high levels of residential mobility, dual residence and all the other features that give social life its fluidity. The point is not that the *miit* identity of individuals and groups is idle or irrelevant to practice, but that it is always contingent upon the discursive context, and is therefore liable to be contested. Even when there is no contestation, it is not because identity has transcended such contexts, but because, *pro tem* and for the purposes to hand, a consensus has been achieved.

In the context of the dispute over Nena, the complex notions about connectedness and the identity of open, yet endogamous, communities go some way to explaining how it is that both sides can claim that they have succeeded the Ontouten. The Telefols claim the area on the grounds of their conquest of the Ontouten and through their descent from captives of the original residents. More recently, they have also advanced arguments based upon the cosmogonic activities of their primordial ancestors. The Wamei Miyans, who deny that the precise area in question was ever annexed by the Telefols, also assert an originary identity with the Ontouten, and point to both their assimilation of the routed *miit* and their cult sacra as grounds for their claim.

The discursive and practical considerations described above, and which underlie the positions of the protagonists, give the fundamental concept of *miit* something of an essentially contested character and underdetermine any definitive answer to the question of which side and which set of individuals exclusively 'owns' Nena. Interestingly, there was a move, late in 1995, on the part of certain influential Wamei leaders, to suggest that the Ontouten *miit* was still in existence, composed of all those on either side who could trace a definite genealogical connection to an Ontouten ancestor. Although, reportedly, there were those on the Telefol side who were ready to consider this suggestion, it did not – initially at least – carry the day, largely because, on both sides, it would have excluded the bulk of the claimants' relatives and co-residents.[13] Nevertheless, the proposal to reconstitute this – bilingual – *miit* as the true 'landowners' was not an outrageous one and was, in fact, no less consonant with the logic of the concept than any other proposal that might have been made. The fundamental problem here, of course, is that a notion which was primarily oriented to making connections, to include people in a context where they were a significant limiting factor, is now being called upon to provide criteria of exclusion.

The predicament facing those in search of a definitive answer to the question of who owns Nena is redolent of certain thought experiments that scholars concerned with personal identity tend to go in for. We might consider a pseudo-Borgesian story that makes the point:

> A Babylonian entrepreneur decided to send employees skilled in detective work to retrace the route of Jason and his ship, the Argo, with a view to collecting relics from the epic journey. Luckily for the entrepreneur, Jason had had occasion, at one point or another, to repair every single component of the Argo, and his employees were able to find every one of the discarded or broken parts. These were carefully shipped back to Babylon, where skilled workers restored each piece and then reassembled the vessel, which the entrepreneur set up in a theme park. Meanwhile, however, a rival entrepreneur had found the ship that Jason and his Argonauts had returned home in, more or less abandoned in an unlikely and remote harbour, and he bought and restored it and set it up in his own theme park. The problem is that both parks advertised their attractions as the authentic, actual 'Argo,' in which Jason had carried out his legendary deeds. The matter is still before the Babylonian courts, and has been for many years, while the judges argue back and forth.

The point of fables like this is to show us that our intuitions about the identity and continuity of socially constituted entities – our doxic responses – can leave us stumped in situations that are actually not so far removed from, and have some obvious kinship with, the normal course of affairs where they work quite well. (This is not, of course, to impugn those intuitions: they do the work they do and there's an end of it.) Perhaps the Mountain Ok disputants over the ownership of Nena are in a comparable situation. Under the circumstances that prevailed fifty years ago the possession of Nena might have been settled, or it might have continued as a border zone which no side could use fully, but this consideration could not indicate how the current issues should be decided.

I hope that this section has highlighted those features of Mountain Ok social organisation that render it difficult to reach any non-arbitrary conclusions about who qualifies as the exclusive 'landowners' of Nena. The aims of this chapter, though, are still nowhere near being attained. For all that I have said so far, the fact that there was initially some agreement about the ownership of Nena, and the distribution of occupation fees, is itself in need of explanation. What have I so far said that might offer an alternative to the obvious explanation – that the Telefols changed their minds, or said they had, under pressure of their desire to get their hands on the royalty money? I hope to provide at least a reasonably plausible response to these questions through a consideration of Mountain Ok conceptions of the nature of the world, of knowledge and of truth.

Knowing and acting in the world

All ethnographers of the Mountain Ok region have been struck by the variation and inventiveness of cult traditions. I have described elsewhere the variation that exists in cult practice and the distribution of secret, sacred knowledge within the western Miyanten with whom I worked in the 1970s (Gardner 1981; 1983). Work among almost all the eastern Miyanten *miit* in the 1990s confirms that it is consistency and doctrinal orthodoxy over any span of social space that is the exception (see also Barth 1987; Brumbaugh 1990).

It is also widely reported by ethnographers of the central and northern Mountain Ok that the Telefolmin have the most comprehensive knowledge and the most powerful cult traditions in the area. Yet, Dan Jorgensen reports, the Ifitaman Telefol elders – the most central and most powerful of all – hold that the indeterminacy of the world transcends human knowledge, including the sacred knowledge of the cult (see especially Jorgensen 1980; 1981b, chapters 4 and 5). Nevertheless, Telefol cult elders are regarded as having profound knowledge of the hidden processes of the world, and organise long complex ritual procedures that draw upon that knowledge to ensure the empowerment and maturation of their youths and the productivity of their subsistence practices. Almost all other groups in the region perform initiation and fertility rituals to the same ends.[14]

Considered in the broadest outline, the cult is an aspect of a lifeworld wherein the permanent and transient structures and processes of existence are aspects of the agency of beings of various sorts. Different forms of bush spirits, ancestral spirits and the spirits of living human beings are not merely part of the furniture of the world, but are the efficient causes of all significant events. As such they are the primary resource for understanding how the world is and how it functions. And just as a person's actions cannot be comprehended at the level of mere behaviour, but must be grasped as an intentional move in a context (this for familiar wink-twitch sorting reasons), understanding the Mountain Ok world amounts to interpreting the intentional states of non-living beings – their desires, jealousies, anger and so on. In short, Mountain Ok peoples dwell in a world that our anthropological forebears would have categorised as animistic. The most significant category of agent for the cult is that of ancestral spirits, especially those whose cosmogonic acts were responsible for the basic features of Mountain Ok life.

It is tempting to view the indeterminacy Jorgensen's Telefol elders referred to as stemming from the difficulty of discovering the fundamental facts, of knowing exactly what happened, rather in the way physical cosmologists are currently unclear about what exactly happened at the beginning of the universe. I would suggest, however, that this epistemological framing of the matter ignores the specificity of agency-based cosmologies of the sort found among the Mountain Ok by casting them as similar to either the physicalist or religious cosmologies with which we are familiar. In both the last two, the world has a standing structure and organisation (either simply in the nature of the principles ordering the world, or because the creator made it so) about which we may have greater or lesser knowledge. In Ok cosmology, I wish to suggest, there are no such inviolable principles, save that all features of the world are direct or indirect aspects of ancestral desires (although, of course, this is not articulated as metaphysical dictum). Under these circumstances, the understanding of the cult elders represents the asymptote of human interpretive capacities, which can never attain closure, just as the doctrinal representation of the *miit* stands as the limit of organisational practice in context where political communities are always open.

First, let me give some ethnographic data which I hope illustrate some of the issues I am trying to make clear. I once collected a detailed account of a named ritual stage from an elder of a lowland Miyan group in the presence of a man who had been through a ritual of the same name in a higher altitude (and more prestigious) cult house. In discussion afterwards the latter felt no inclination to disparage the very different and much simpler procedures enacted in the lowlands cult (although such disparagement did happen quite often). When I asked him what he made of the differences he shrugged his shoulders: 'What can be said? It is true that in the east, in the original homelands of the Miyanten, our ancestors showed us how to perform the rituals we still perform, and these are different from the ones performed in the lowlands. But look at this [lowlands] elder. His taro grows well, he kills wild pigs and cassowaries and his skin is strong. His ancestors look after him just as our ancestors look after us.' On another occasion, after several days of interviews with knowledgeable men on the subject of taro fertility, I reported, in an amused tone, a dream I had had to my informants. In it a wizened, venerable Miyan elder, whose face I did not recognise, and whose hair was bound up in the traditional manner (except that it was unusually phallic), fetched me and took me to a garden with a promise that he would show me what really made the taro grow. When we arrived at the garden it had been cleared of undergrowth (but in the usual way the largest trees were still standing), and holes had been made in preparation for taro setts. The old man approached one of the holes and lay on his back next to it. He then inserted his top-knot into the hole and began moving his head up and down. When I recounted this dream, the senior man present instantly urged me to lower my voice lest any uninitiated person hear it. He said that the old man must have been his mother's father, a legendary cult elder, who knew that I was interested in the *miit* of taro fertility. 'Dreams,' he said, 'are the radios of the ancestors.' Astonished, I asked him if any operation of the sort shown in the dream had in fact been part of taro fertility rituals. He replied that it had not, but that did not mean that it might not be efficacious. Other examples of what strikes one as an open, flexible orientation to ritual could be given (and I have done so elsewhere). Fredrik Barth recorded nine innovations in ritual among the Baktaman, three of them occurring during his period in the field. Some of them involved the abrogation of taboos and others the introduction of new temple sacra (1975:Chap. 27). He also found that the marsupial featuring in one of the most sacred myths was not the one actually used in temple offerings. He was told, in explanation, that the first marsupial had previously been 'strong' in producing taro but had since proved less effective than the new one (94). Barth remarks that these innovations indicate the pragmatic dimension of sacred ritual. Jorgensen, summarising a subtle account of Telefol secrecy and knowledge suggests: 'The aim of Telefol religion is less to produce a systematic and air-tight view of the world than it is to produce men who are capable of dealing with [it]' (1981b:490).

The Mountain Ok world, as experienced, constitutes the revealed – but not necessarily settled – preferences of the ancestors. Accordingly, innovations, in so far as they are efficacious, likewise index the desires of the ancestors, since nothing that succeeds can do so of itself. There simply is no non-intentional stuff that acts in the world through its own powers. It is important to stress this aspect of the life-world we are considering: no feature of the world has any efficacy in itself, at least none that anyone knows or cares about. (This marks out the kind of life-world we are discussing from other sorts of religious worlds: after all there is no contradiction in being a Christian and a chemist.)

Understanding and acting effectively in such a world is a matter of interpreting things properly rather than knowing how it works: in terms of a familiar dichotomy, it is a matter of *verstehen* rather than *erklären*. Of course, it would probably be better to characterise the Ok world as providing no grounds for this distinction even to be drawn. It is still more accurate, however, to see the interpretive mode as fundamental, since there is simply no space for the non-interpretive to occupy. Hence, all understanding is a question of grasping what things mean, in the manner of interacting with fellow human beings: understanding the point of their actions, rationalising their projects in terms of what they desire and what they believe. Human projects are always moves in a conversation – the conversation of the world – which are based on interpretations aimed at understanding what more powerful interlocutors want and do, for it is only by doing so that human projects can be realised effectively. The events of the world, small and large, are expressive of the intentional states of beings, and the world of 'things', accordingly, has all the marks of the mental that is characteristic of the world of persons.

Even those routines of life that humans themselves carry out depend for their success upon the ancestors or some other category of agent. Accordingly, the characteristics of intentional understanding, which always remains within the hermeneutic spiral, obtain in all realms of life. For the Mountain Ok, it is interpretation all the way down. The efficacy of initiation and fertility rituals depends upon the elders producing and using not symbols of ancestral power, but samples of it (Gardner 1983). What makes the difference between leaders and the led, between cult elders and those they initiate, is the ancestral grace they enjoy; that is, the amount of power they can draw upon. This, in turn, is an aspect of the social relationships between living and non-living agents; that is, they are interpersonal – usually kin – relations, and must be attended to with the same care and attention as relations between the living are. Everyone, though, directly or indirectly, depends upon relations with those beings whose psychological states are expressed in the flow of the world.

While, as among ourselves, most interpretations are done on the run – are, we might say, immanent in the flow of life itself – it is in the nature of things that they are defeasible. Just as a settled interpretation – of, say, a political act – must be revised when we learn something new about its context (that, for example, the MP was working for a foreign power), for Mountain Ok peoples, the world itself may call an interpersonal interpretation into question. And, given the mutually constitutive nature of the link between context and meaning in relation to actions, there is an inevitably holistic tendency to the interpretive enterprise, which in agency-based cosmologies implicates the world as a whole. As the elders repeat again and again to their charges in the initiation rituals, something as trivial as the theft of produce from a garden, or the repetition of a cult secret to someone not permitted to hear it, can precipitate a disaster on a major scale, as when a community's taro crop fails because an angry ancestor has willed or allowed it to happen. Paradoxically, the ever-present necessity of interpretation – what we might characterise as the 'non-foundationalist' nature of Mountain Ok cosmologies – gives those who live in such a world a remarkably empirical outlook (see Bateson 1958:230-2 for a discussion of the Iatmul spirit of intellectual inquiry). If events are always aspects of the actions of agents, and thus must be read hermeneutically and defeasibly, then this makes what happens more complex and more worthy of attention than any purely physical construction of it would.

This non-foundationalism, and the hermeneutic-empirical outlook it provokes, extends to the self. Since Descartes many Westerners have come to regard the self as the realm paradigmatic of certainty. Against this we can cite the sorts of situation where we must observe ourselves to discover some aspect of, for example, our knowledge, as when I recite a poem to see if I still know it or write down a word to see if I know how to spell it; or our character, as when I discover that I must revise my assessment of my bravery after experiencing a novel form of danger; or our beliefs, as when I discover that I do not, after all, believe that following God's teachings can never lead us into great evil. Mountain Ok understandings of self seem to me to be more firmly grounded in experiences of this second sort. People may really wonder, for example, whether their resentment of another has led him or her to become seriously ill. These considerations also help to explain something that initially puzzled me for some time: this is what I saw as the unusual moral valency that attends the accusation that someone is speaking untruthfully. For us, by and large, to be charged as a liar or as having got things wrong unintentionally, is to have one's standing quite seriously impugned, but I was always struck by how relaxed Mountain Ok people are in making and being the subject of such charges. I now believe that this is explained by an appreciation of how ubiquitous the need for interpretation is in a life-world of the sort they inhabit. For them, speaking untruths is largely a matter of misinterpretation, of the sort that I acknowledge when I come to see that a man's exuberance is a product of great nervousness after having previously seen it as an aspect of his supreme confidence.

In such a world there can be no question of humans taking or having power in themselves: power cannot be taken, only given. This includes the power to state how things are or were – to utter truths. But since how things were or are is an aspect of the will of the ancestors, and since even the most secret knowledge is liable to be found unreliable, 'correct' or 'true' can only mean 'endorsed by the ancestors'.[15] Ancestral endorsement of an interpretation manifests itself in the efficacy of actions it warrants; likewise, interpretations that the ancestors do not endorse reveal themselves in the failure or ineffectiveness of actions they entail.

Under the conditions I am trying to characterise, it is hard to draw distinctions between possessing and owning, between having power over and having rights over a resource, distinctions that are crucial to our category of theft, for example. This is because, as I have already said, in an agency-based cosmology like that of the Mountain Ok, power can only be given, not taken. Consequently, continuity of possession ('ownership') of resources (land, tree-crops and so on) is a matter of sheer narrative history, and the identity of the possessors is constituted through their succession. For if one group of possessors is replaced by another (through conquest, for example), then the continuity is established by the very fact of success, because this bespeaks the acquiescence and ongoing support of the relevant ancestral beings, which is the only form that legitimacy can possibly take. (That this is the issue is suggested by the way those who take land by conquest assert their continuity with the previous possessors, and victors frequently keep alive some of those knowledgeable about an area to ensure that they behave appropriately *vis-à-vis* the spirits abroad in it.) If this indicates a certain capriciousness on the part of these superordinate agents, then so does much else about the vicissitudes of human life. (One other example: the Miyan elder responsible for taro fertility, whose powers are vouchsafed by a spiritual counterpart, is required to restrict his movements lest his travels should lead him somewhere more pleasing to the spirit upon whom his vital function depends.)

If, as has often been suggested, Western thinkers since the Greeks have responded to the radical sceptic's challenge with an attempt (invariably vain) to build an epistemologically bullet-proof framework, the world of the Mountain Ok suggests a response to epistemic indeterminacy that is more like a shrug of the shoulders and an observation that *c'est la vie.*

Conclusion

I hope it is by now obvious where I see this as going. Both sides rest part of their claims to Nena on their interpretations of how things sit with the ancestors, and the power to possess the area that has been devolved upon them. The Telefol change of heart about the area is not necessarily to be viewed as a direct expression of a lust for money (and even if money were the immediate object of their manoeuvres, I would argue that, in the absence of an account of how money is construed, this is still an unacceptably thin characterisation of their actions). The claims the Telefols are making now are no less reasonable, and no less continuous with their previous practices, than those of the Miyanten.

I should like to make it clear that I am not denying altogether that there is a link between what the Telefols are saying and what they desire, merely that the direct, rationalising connection posited by the 'obvious' explanation is incorrect. Even Western ethical systems acknowledge that the passions interact in complex ways with our cognitive states, and that there different ways in which, say, a desire can bring about a given outcome. There are, that is to say, what philosophers sometimes call 'deviant causal chains.' (As when a man's hatred of another leads him into a reverie of revenge that distracts him from his driving and causes an accident that kills the object of his hatred. The hatred caused the person's death, but it wasn't the basis for what occurred as it would have to have been if we entertained the question whether this was a murder.) A deviant chain, I would be prepared to concede, may be linking the desires of the Telefols and their actions in claiming Nena as their own. But even this concession seems to me to be unnecessary in view of the descriptions offered above. A more straightforward, if less obvious, explanation of their conviction, and the supersession of their previous position, comes from seeing them as perpetually involved in interpreting the nature and extent of the powers that their superordinate ancestors – the fundamental processes of the world itself – have bestowed upon them. In such a context, the power to state particular truths is itself in question, and the veracity of any account can only be vouchsafed by the this-worldly success of those who act upon it.

Notes

1. The dispute was apparently resolved, at least to the point where compensation could be paid by the developer, in July 1997.
2. This is an exercise in interpretation at a rather general, cultural level. A more fine-grained analysis would take account of the variety of backgrounds of the principals (some of whom have spent many years in high-level administrative positions, university training, experience in the mining industry, and so on). The 'obvious' explanation, I would concede, might account for the behaviour of some of the protagonists, at least at a rather crude level. My efforts in this chapter are directed only to subverting its appeal as a general characterisation of motivational currents at play.
3. Both Morren and I have, hitherto, used the official name 'Mi(y)anmin' to refer to this population. The suffix *min,* however, is Telefol, and the administration's adoption of this form reflects the history of contact in the area. Given current sensibilities, I here use the Miyan form *ten* (which also means 'people').
4. At the time of the studies referred to here, the developer was Highlands Gold Pty Ltd. Currently, after a takeover and redistribution of its assets, the developer became Highlands Pacific Pty Ltd.
5. Members of two other groups, the Iwam of the May River and the Owininga, have suggested that they are descendants of the Ontouten driven out of the Nena by the Telefols.
6. The myths suggest that the ancestor of the Nenataman Telefols secreted mineral wealth in the Nena area (see Jorgensen (this vol.) for details).
7. In 1995, both George Morren and I collected song texts and wordlists in this tongue. Dan Jorgensen speculates that it is related to the language of the Paiyamo (pers. comm. See also Jorgensen 1996).
8. Among the Miyanten, a penumbra of vague, rarely articulated beliefs surround these core tenets, for example that the heart is also of paternal

origin, but, so far as I am aware, these play little role in standard discourse on such matters

9. In view of the similarities across the area, it is a plausible hypothesis that the differences in population densities within it are related in some way to the variation in the health hazards of different ecological zones.

10. The claim is, of course, conditional upon the state of pre-existing relations between migrants and their hosts. In practice, though, only a state of enmity precludes such residential shifts.

11. It is interesting to note that there is a significant pattern wherein immigrants, or their sons, become the leaders of a community.

12. *Hon* is also used in adjectival constructions to mean 'barren'.

13. But this appeared to be the basis of the settlement achieved in July 1997. It was reported (in the *National*) that Telefol Ontouten and Miyan Ontouten shared, in equal proportions, the first compensation payment made by the developer.

14. The advent of Christianity has had dramatic effects upon these ritual complexes, but they have not in any simple sense been superseded. On the contrary, the mythopoeic integration of cultic and Christian ideas has produced striking continuities with previous orientations to the world that are, moreover, significant to the interpretation of mineral deposits (Jorgensen 1990 and this vol.).

15. I am not suggesting that the Mountain Ok believe that the ancestors can or do modify history.

6

LAND, STORIES AND RESOURCES: SOME IMPACTS OF LARGE-SCALE RESOURCE EXPLOITATION ON ONABASULU LIFEWORLDS

T.M. Ernst

In this chapter I explore a particular dimension of apparent change in Onabasulu lifeworlds. It is specifically concerned with aspects of their public and intersubjective reflections on sociality and their place in the world. The change, as I describe it in the first instance, is one of degree. That is, it involves what is, at first consideration, a remarkable continuity of cosmological precepts which maintain relevance in greatly altered historical circumstances. However, as Sahlins said early in his historical studies of Hawai'i:

> Yet only in certain respects and appearances is it true that the more it changes the more it remains the same. A view of history that was content to see in the formation of classes and the state ... merely the reproduction of the traditional structure would arbitrarily limit the powers of anthropological understanding. Cultural theory has no need to be embarrassed in the face of structural change ... [Sahlins 1981:31]

Within the Onabasulu context of apparent cultural continuity, there are processes of change. They are manifest in what, at first sight, seems an acute aspect of the continuity itself. The apparent continuity is found in discursive practice that treats explicitly aspects of the sociocultural world which, in the recent past, were mainly implicit. It looks like a cultural 'self-awareness' in which there is a willingness to develop 'exegeses' of aspects of the culture. Similar processes may be relatively widespread in Papua New Guinea in the context of modernity. There are indications of it in what Stürzenhofecker describes among some Duna speakers who are also on the immediate periphery of a large-scale resource development (Stürzenhofecker 1994). In the Waghi, O'Hanlon (1995) cogently discusses the designs on the 'reinvented' shields in the context of a new resurgence in warfare. He finds a kind of discursive reflexivity in which the visual art by the Waghi people is discussed and objectified. This was absent in his earlier ethnographic experience. There are parallels here with Onabasulu discussions of their previously contingent or implicit social categories as entities. Certain categories of social importance, in particular ethnic categories and kin groups, are constituted as objects in ways that appear different from situation in the early 1970s.

In Onabasulu, there has been a marked tendency toward the *entification*[1] of social and ethnic categories. It is a process of the making of 'entities,' or things from what have been either implicit or contingent categories. It involves a concentration on their *existence* rather than their effects, qualities and relational nature (that is, on the categories 'being' instead of their continuous 'becoming'). This process of entification was striking on a recent visit to the Onabasulu after a long absence.

I prefer the term entification to reification for two reasons. First, entification places an emphasis on the importance of the being of the object as 'attended to.' The 'thingness' is not just taken for granted, as it is in the case of those usual objects of reification, social typifications. The objects of entification that will be discussed result from a specific attention to their 'thingness'. Their entivity is a result of a particular sort of attention which focuses on their being.[2] The second is that the term 'reification', in social science usage over the past three decades particularly, carries with it a notion that the attribution of a facticity and externality to the reified object is ontologically inappropriate. Reification connotes a sort of false consciousness.

What is important in the ethnographic cases cited above as well as in Onabasulu is that the objects developing an externality were described as implicit in previous accounts of social life in these populations. There is a certain gratification for anthropologists in the process, as it seems to validate and elaborate earlier work in which similar objects were constructed by the analysts' reflection on implicit and unstated categories. The difference is that the objects are now conceptually distanced from peoples' social life through a new form of attention paid to them. This change is one that is always considered a possibility from the perspective of a constitutive phenomenology of social life.[3] What I establish in this chapter are the specific conditions under which this form of reflection on social life has become *a regular and continuing* part of *general* Onabasulu discussion. This is opposed to the odd cases that analysts are so dependent

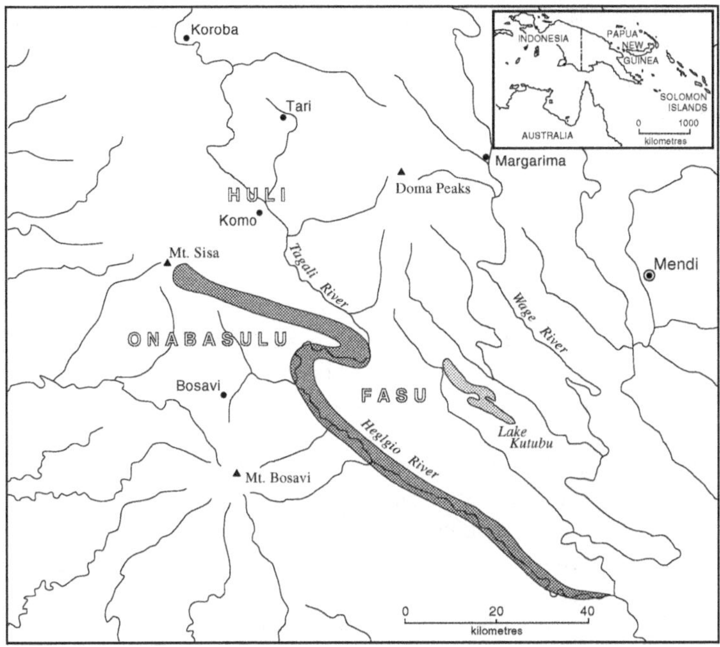

Map 6: The Onabasulu-Huli-Ipili region of the Papua New Guinea Highlands, together with the Onabasulu mythic snake. (Adapted from Biersack 1995a:2 and Frankel 1986.)

upon from specific persons ('informants') at irregular times. I locate these conditions in the specific recent historic circumstances surrounding oil extraction in a nearby location and the inclusion of Onabasulu land in an oil exploration (and the development of petroleum) license held by Chevron Niugini Ltd.[4] An important part of the specific conditions under which this formerly infrequent manner of 'attending to' social categories has become common (and thereby socially effective) is the way in which strong interaction of 'customary' social practice with legal institutions has been precipitated by the concerns of capital and the Papua New Guinea state.

It is possible entification in such a context may have far reaching and potentially transformative effects.[5] It is the process of entification upon which I focus. I do so in the context of three stories I collected in 1996. All three stories are abbreviated versions of material recorded in Walagu 'village' in January 1996. The first concerns geographic features, the second concerns an autochthonous woman, an originary figure named Duduma and the third traces kin relations and control of land. They were told by several people, all negotiating the basic narratives, and then recorded on audio cassettes in informal public gatherings in Onabasulu and Pidgin by Agibe Fua, Yeya Deba and Malime Deba.

Story one: Marks and Faiyaninaro, the big snake

Now we have land group certificates. These certificates come from the National [government] in Moresby. On these we put seventeen clans that are inside. So now, if other people try to come inside and say the land is theirs, they are talking deceptively. Our belief is that the seventeen lines are authentic. Now if some company comes inside and wishes to work, we have certificates. Now that we have Land Certificates, we no longer have worries. Suppose a company wants to mine gold or drill for oil and gas, we are the true owners of the earth (*papa bilong graun* in Tok Pisin).

We have a big mark. On the Fasu side, we have the large river, its name is Mbubi [the Onabasulu name for the Hegigio, and not to be confused with the Mbubi River which appears on map 6 to the West of the Hegigio]. Cassowaries cannot cross it, snakes cannot cross it. People, too, cannot easily cross it.

On the side of the Huli we have a large wall, a big mountain. This mountain we call Folola. Planes find it difficult to get up to get over this mountain. On the two sides of this mark there are different stories. On our side we have sago and many kinds of snakes that can kill people. On the side called Huli, they are different. They have grass (*kunai*), and they build a different sort of house. We build long houses on hills or ridges. We have the mark, this mountain Folola. It finishes in the mountain Aowaga and finishes at Mbubi. On the top is Haliago (Mount Sisa, Huli term), and it finishes at Juha. One side is the side of the Onabasulu, one side is the side of the Huli. Huli should not come into this side, the side of the Onabasulu. Onabasulu should not go into the side of the Huli. These mountains are our border.

There is a road. There is talk of a new road. We have a road from before. Huli sometimes come with weapons and kill our pigs and rape our young women.

We have a story about Fofola and Aowaga. There is a Big Snake. The snake's name is Faiyaninaro. The tail of this snake goes down past Auwaga, down the Kikori River. The head finishes at Juha. The grease

> of this snake is found as oil. The blood of the snake is gas [one person suggested it was its brains that were gas]. It urinates petrol. This is an old story, the story of the snake. (It was part of our male initiation.) It is a true story, not like the stories told by some Huli of a snake, the tail of which is on the Huli side and the head of which is on the Onabasulu side. The snake sleeps. Its tail is down the Kikori, its head is at Juha. White people do not know this story. People here know it. It is why we believe that most of the oil and gas in the region are here. It is our belief that Chevron knows that the big deposits of oil are here, and will be worked when Kutubu fields are finished.[6]

While the general ethnic configuration of the region between Mount Bosavi and Mount Sisa has recently changed significantly, I shall concentrate on Onabasulu-Huli relations and their implications in this chapter.

Resosis, geography and ethnic reconfiguration

It is not misleading to say that presently, images of the Huli dominate much Onabasulu public discourse. There are sermons in church about Huli and the problems of their '*raskol*' gangs, there are constant references in everyday conversation about Huli, and the Huli are seen as one of the greatest threats to the Onabasulu attainment of wealth through *resosis* (resources – especially oil, although some Onabasulu and Kaluli hold a firm belief in the presence of gold in Mount Bosavi).

Earlier relations with the Huli were reserved, with little if any raiding, and based primarily on trade relationships. By 1996, the situation had become more fraught and in many ways, better defined than in the 1970s, in part (although not exclusively) as a result of speculation by peoples in the area on the location and control of access to '*resosis*'.

When, in late 1969, I began research in the Bosavi region, with a category of people who had a 'discrete language', who came to me 'pre-known' as the Onabasulu (or Onanafi in the Etoro language) from the very recent and very high quality ethnographic work of two close colleagues, I was entering a relatively straightforward situation by New Guinea standards. It was a named language corresponding with a named category of people – defined in their interaction with other such groups around them. As is the usual ethnographic experience, the category dissolved rapidly before my eyes. I began thinking I was located on a thick boundary line – drawn with a blunt pencil, between the Kaluli and the Etoro. To be sure, it had its own distinct language – but it was only one (albeit the dominant one as you approach the centre of Onabasulu settlements) of the languages used in everyday life in 'Onabasulu' long houses. Very early in my research then, I continually asked questions about Onabasulu, where they started and stopped, what were their diagnostic characteristics – all, of course, to no avail. The term Onabasulu was seldom used – usually only in response to my inquiries.

What was clear were the dominant principles of differentiation at the edges. They included kinship/affinity, trade and violent hostility. It is where there are intense nodes of kinship/affinity that the edges are the most poorly realised and blurred (for example, Onabasulu-Etoro and Onabasulu-Kaluli historically). They become more distinct between groups where the dominant mode of interaction is something other than kinship. Trade is one such mode. Historically, this was the case with Onabasulu-Huli relations, The most effective way of creating distinction, up to the 1960s, however, was bellicosity: traditional relations of violent enmity (such as those between the Onabasulu and Fasu).

In discussing ethnicity on the Great Papuan Plateau, it must be remembered that although distinctions (and resemblances) were important, boundaries and groups historically were not. To read in to the categories of participation, sociality and ethnicity (sameness and difference) the radically separate bounded cultures sometimes described usually does the ethnography, particularly the earlier ethnography, a serious injustice. The world of bounded categories of ethnic identity seems commonplace to us because it resonates with modern ideologies of national cultures, particularly those of Europe in the 20th century (and other nationalist projects derived from them, particularly after World War II). But it does not seem to fit well with understandings in sociocultural forms on the island of New Guinea. However, there has been a move, at least for certain contexts, to try to develop the clear boundaries so useful (if ultimately definitionally slippery when questioned) in particular in recent expressions of a politics of 'cultural identity.' The words 'border,' '*mak*' and '*banis*' have become a part of common usage in discussion (just as words such as 'royalties,' 'equity,' 'shares' and 'management structure' receive moderate to occasional use).

The increasing perceived (and actual) heavy increase in belligerence between Onabasulu and Huli fits into the creation of sharp difference. So too does the story of the snake that both defines the boundary between Huli and Onabasulu and accounts for why the story is important.[7] Interestingly, though it reifies the category 'Huli' (*Disie*), the story also – and this is more important for considering conceptual change – gives *entivity* to the category 'Onabasulu'. This change is immediately noticeable when, upon landing at the small airstrip near Walagu, you can see a short distance away the new subdistrict '*Onobasulu*[8] Mini Health Centre' duly named and labelled as such and ceremoniously opened by national government officials in November 1995.

The Death of Duduma and the origin of clans

Story two: the story of Duduma

This is the story of Duduma, now said by some to be the most important story, for those who 'know' it controls Onabasulu land.[9]

> Duduma was a large autochthonous woman in the time when no kin groups or places had names. She lived in the head waters of the Kadi River near a place called Maliya. Two men, one of whom was her husband Wafesisila and the other a 'good man' named Desio talked about and decided to kill her treacherously. Desio talked to her while Wafesisila worked his way behind her. He grabbed her and held her while Desio killed her. The place where she was killed was a small body of water Ibisugoana. The meaning of the name is that the blood of Duduma went down to here. This is between Kebi and Hwogosie. The big river here is the Kadi River, and she was cut up on the Kebi side, close to the Kadi. Desio cut her up at Maliya. It is also *Mi*, or the centre [*mi* also means seventeen and nose].
>
> Her blood became the paint that colours the megapode's (*aro*) legs red. Her liver became the Sewa River. Her body was carried around the entire Onabasulu country. Where she was killed, sago grubs (*feleli*) came up in profusion, and a type of marita pandanus called *mimaro* grew.
>
> She was carried around the country and her parts were distributed among the seventeen true Onabasulu *mosomu*. They received their names from these. [Now the word *mosomu* is glossed in Tok Pisin and English by the Onabasulu as 'clan'; formerly the gloss used was '*lain*'].

> These are enumerated. Some examples include Kimise being given her bones (*kiwi*) (Kinido was also said to be given the bones – it is now spoken of sometimes as a subclan of Sogaise, which received the leg[s]. Kinido was formed by fission from Sogaise, and established by a man named Agiba in the early 1970's), Hanoro being given her body fluids (*hano*), Gunigamo being given her head. Kebi struck and killed her, cut and cooked her, and has a set of names from the process; including Kebi, to touch or hit her, Kebi Abane from the cooking stones, Kebi Fuagba, from the sound when she fell (*fu*), etc. [The 'complete' list of seventeen clans was not recounted at this time, but there was an assertion that there are seventeen incorporated land groups.]

Maliya is an origin point. It is the centre of the world. Near there, on a hill on the Hwogosie side of the river, where she was first grabbed and hit and then pulled down to where she was killed and cut up, there is said to be a small cave that has a stone dish in it with tree oil in the dish. Floating in the dish is a device made of sticks in the form of a cross with a third stick perpendicular to the others. A specialist from Hwogosie can look at this device, representing the four directions of the world, to tell if there is a maldistribution of people – people coming back to the origin point. Slight problems in the distribution of people result in earth tremors (*heleli*). On the Kebi side there is said to be a tree that grows out of the ground and back into it in the form of an arch. A Kebi specialist can read the growth of shoots off the trunk as signs of the state of the ground.

Background

This is a central Onabasulu myth, and much can be gleaned from it. It is about fragmentation and dispersal, as a result of an act of treacherous violence that is cosmogonic. It is from this that sociality – distinctions of clans, marriage as a control of dispersal and so on – develops. But this is a story that previously received little outside circulation. Now a number of Onabasulu want it to be widely known. It is seen as central to an Onabasulu identity, and has much in common in terms of political functions with story one.

This short section, however, will just indicate where it fits and give some history to the way people may now want it widely known. The place where Duduma was killed is mentioned, without the myth, in Schieffelin's discussion of Papuan Plateau accounts of Jack

Hides 'exploratory patrol' in the 1930s (in Schieffelin and Crittenden 1991).

The place where Duduma was killed is referred to in Schieffelin (1991:67) as 'also at another (lesser) origin spot (also called 'Malaiya') near the headwaters of the Kadi River among the Eastern Onabasulu'.

> In fact, the people of Ogesiye hiding in the forest nearby had been interrupted in the midst of playing out a desperate drama ... The Ogesiye community was custodian of the second Origin Spot (Malaiya #2) at the headwaters of the Kadi, and they had been especially involved in the controversy that accompanied the first appearance of bush knives and axes in the area. The appearance of Hides and O'Malley now confirmed what they had feared: these objects were the very ones depicted on the cliffs, and now, as some had predicted, their 'owners' were returning from the Origin Time. [Schieffelin 1991:80]

Schieffelin goes on (note that Wabowe (Wabue) appears in story three):

> Wabowe of clan Keibi, who was about eighteen at the time, recalls seeing Hides from the Ogesiye veranda just as the patrol emerged from the forest. He was motioning for people to sit down and not run away. But everyone was panic stricken. The light-skinned man they saw was wearing a dark shirt, and they thought it was part of his skin. 'If he had been wearing a loincloth, it would have been all right, we wouldn't have feared,' Wabowe [Wabue]declared. 'But when we saw his clothes, we didn't know what it was. Was this a spirit person or what.' ...
>
> It was during this flight that Hides captured the old woman, Hayoba and after calming her, presented her with cloth, beads and an axe as gifts to her people. Not surprisingly, when she brought them to the people hiding in the forest, they were thrown into even greater consternation. The spectre of the world collapsing to its origin point loomed over the discussion that Wabowe remembered: 'These are taboo things, not for human beings to touch! If you take them, will their owners not follow them? Give them back so they will go away'. [Schieffelin 1991:81]

In 1970, a petroleum exploration company which was a subsidiary of the Bendix Corporation carried out seismic surveys on the Papuan Plateau. They set up a camp at the junction of Kulu Creek and the Libano River, in a 'grey area' between Onabasulu and Kaluli.

Many Onabasulu worked for the company, laying out the charges and guiding helicopter pilots. The line move up the Libano and Kadi River, alarmingly close to Maliya. I was told only of the general significance of Maliya at the time (with less detail than Schieffelin was given later), but it caused alarm. In reflecting on the petroleum prospecting twenty-five years later, various Onabasulu draw two conclusions, based on the geography of the two stories above. The first is that the disposition of the snake Faiyaninaro and the oil search provide mutually corroborating evidence of oil deposits in Onabasulu territory. The second, as I was repeatedly told, is that it was important that the first phase of petroleum extraction occurred at Kutubu, for if a settlement as large as the place Moro, on the northern end of the lake, especially during the construction phase in 1992, when it resembled a small city, had been built at or near Maliya, the earth would have been irrevocably disturbed or destroyed (according to the importance of story 2).

The seventeen clans and the Onabasulu

The Onabasulu term for kinship group and house group is *mosomu*. The common term consists of layered and contextual meanings and carries no notion of a 'total' congruity between local groupings, kin groups and long house groups. In fact, normative residence arrangements take cognisance of aspects of social organisation that preclude the complete localisation of kin groups. I chose to call these kin categories lineages some time ago for reasons that do not seem as compelling now. I continue to call them lineages, however, just because I avoid the word clan by doing so.

With regard to 'real property', Onabasulu lineages are (or were) at best weak 'corporations' if corporations at all. At least this is so once any members of a lineage establish a long house on its territory. Any 'exclusive' rights in use of land for gardening are diffuse and at best nominal. Lineages have no real control over hunted or gathered 'bush' products, and sago holdings are handled by a separate set of kin-based considerations focusing on individuals, their kin networks, and a history of bestowals for a variety of reasons.

But there are territorial and other connections (mainly men's views on their role in the disposal of women in marriage and the all important anticipated role in the relationship to *resosis* – for example, oil, timber, gold – all very much a commodities futures situation now).

The membership of Onabasulu lineages was small in the 1970s and remains small in the 1990s – seldom more than ten adult men. Lineages have genealogical 'edges' past which members are included only with extreme difficulty. There is, in this regard, a potential for a form of segmentation or fissioning. The stipulation of this boundary is implicit and indirect and is ultimately related to notions of relatedness rather than some notion of lineage structure. People do not trust distant kinspersons, even agnates. Particularly, a man does not trust a kinsman past three degrees of collaterality. This is part of the circumstances that create structurally the possibilities of lineage segmentation or fissioning at points in time.

But the fissioning is never merely a structural consequence of genealogical aging. It is most usually *intended* or possibly (rarely) the result of demographic contingencies, or what is, from a male perspective, the matrimonial recalcitrance of particular women. In any case, whether solidarity is restored, fission occurs, or some form of segmentation is continued (thus leaving open all possibilities) is the result of the histories of the intentional actions of men and women in these small groups. However, with the greater public reflection on *mosomu* and consequent entification of the general concept 'clan', the primary location of segmentation and fission may be moving from practical sociality to discursive practice.

There are a fixed number of clans now. At least, there is said to be a fixed number of clans that are, under Papua New Guinea law, Incorporated Land Groups. These ILGs are presently beloved by government and companies alike as an administrative tool based in local custom (see Sagir, this vol.). Chevron, in 1994, using the same techniques developed at Kutubu, were instrumental in the incorporation of Onabasulu clans as ILGs.

Why seventeen? Probably because it is the centre, the nose, the only number in the Onabasulu counting system that is unpaired, and therefore appropriate to the cosmogonic myth of Duduma. The 'seventeen clans' fixes an Onabasulu identity, but does so by providing, 'in law' a fixed number of incorporated groups.

The man with a tail and the fate of the land of Ole

The story of Wafoale, the last man in the line Ole

Ole is a clan name. Ole was one group. The line died. One man was named Wafoale. He had a tail. To sit down, he had to dig a hole in the ground for his tail to go into. His tail bone went down so he could sit down on the ground. He was a friend of one Sabiasulu man. His name was Gaiyuba. Wafoale gave his land to Gaiyuba to look after, for Ole died out. Gaiyuba died and gave the land to Haiba. Haiba looked after the ground and when he died, he gave the ground to Wabowe. Haiba told Wabowe the place was named Yabolo, Nuguli, Isedo and Wabido. Wabowe had no male children. He talked to Yeya Deba, 'I have no male children. Your two sons, Gobi and Malime should look after the land. Take this ground. Do not give it to other men. When you two die, you must give it to another man.'

This story is important, for if a company comes and finds resources, the owners of the ground are known and the story is taped.

Ole: contestation of control of a dead line

Bruce Knauft has written in a piece about Kutubu Petroleum called 'Like Money You See In a Dream,'

> In a postmodern era, journeys of exploration don't end; they ratchet their ironies to a higher scale. These ironies are not just discursive, epistemological, or limited to a world of tropes, they have enormous impact on people's lives. [Knauft 1996:95]

If so, then Wabue was there for the complete ratcheting. He saw Hides, the Kutubu Project, the coming of the colonial/postcolonial state and the incorporation of *mosomu*. Through it all, he plays a particularly illustrative role. Schieffelin calls him 'Wabowe of clan Keibi'. Yet Wabue was, until his death around 1993, a senior member of Sabiasulu *mosomu*. What he had in common with Kebi is that in the Duduma story, they were implied as married to Hwogosie. So was Wabue. Then, he figures in the third story as a central player in the organisation of what is actually a dead 'clan'. Although Ole is a lineage that has died out, the story related above recounts a version of how its land is to be controlled. This was formerly not of overwhelming importance, but with the process of the legal incorporation of land owner groups, it is no longer just a problem to be worked out in everyday social activity and contested in the context of specific projects. So the new *mosomu* chartered with this story takes the name of an extinct one. It is legal to do so. The Manual of Laws and Procedures for Land Group Incorporation, drawn up for Chevron Niugini says, in the section on the contents of constitutions,

> (a) *name of group* This is something for each group to decide. All that the Act says is that the name should not be 'undesirable', and that it should end with the words 'Land Group (Incorporated)'. [Fingleton n.d., [1993]:6]

Kinship and lineage membership validate speaking positions in many Onabasulu forums. Who is told what, when and where, and just as importantly who is *not* told what when and where is important in the ways Onabasulu talk about kinship and much else. Genealogies are always provided in this context, and as I pointed out earlier, I was seldom the only audience member when eliciting material on who was related to whom. It is important to realise that *all* genealogies are in some general sense pedigrees in John Barnes' terminology (1967). It could only be otherwise if it were possible to remove the same ideal observer completely from the social world, which is, of course, as impossible as it is undesirable.

I offer the following letter that was dictated to me in Tok Pisin and immediately rendered into English by me. It was a collaborative effort of several of the involved parties on 10 January 1996:

> To the Registrar of Incorporated Land Groups.
> Dear Sir,
> There are some problems on the Certificate of Registration of:
> The OLE LAND GROUP INC.
> Mt Bosavi, Walagu Village SHP
> Mr Malime Deba, of this group, says that his name should replace that of Maiyowa Yabe as the Chairman of this group. His contention is supported by Mr Yeya Deba of the Dispute Settlement Authority of this group. I have been unable to speak with Mr Maiyowa Yabe [I suggested it was necessary to add this].
> The changes would then be:
> Chairman: Malime Deba replacing Maiyowa Yabe
> Committee Member: Anton Malime replacing Gaima Maiyowa.
> While I do not know the circumstances of the registration in the first instance, I can attest that Malime Deba is a member of this lineage, and this is apparent in genealogies I collected in the 1970's.
> I was asked to write this letter by Yeya Deba and Malime Deba, and their names are attached

T.M. Ernst
Anthropology
CSU-Murray
Albury NSW 2640
X
MALIME TEBA Yeya Deba (his mark)

I suggested that sending it might not be the best course of action.

While visiting the Onabasulu, I told them I was back for only a short time to straighten out the translations of some material I had collected a long time ago, mainly stories. They thought I should record who were the members of what kin groups and why. This became a full time job, and genealogical details of Gunigamo *mosomu* were being provided even as I boarded a light aircraft at Walagu airstrip. And they were elaborated by a high school student flying to school in Tari on the same plane. He also pointed out various important features including the location of Maliya as we flew over them.

The material is complex and to go into its implications and relate them to my earlier material on Onabasulu organisation is beyond the scope of purpose of this chapter. But several important points emerge from it.

First, I was asked by men if they could 'write their sisters' into their clans. I said that 'they are your groups, write in who you will.' They then 'wrote in' sisters and sisters' children and wives in some instances. I am reminded of what Lévi-Strauss wrote some time ago (his lectures on 'Melanesian Problems'; 1978-79):

> It is possible to formulate matters in another fashion: unlike unilineal systems, which are based on a clear distinction between parallel relatives and cross relatives, New Guinea systems transpose this distinction into the very midst of the category of cross-relatives, who are treated, by virtue of rules inherent in the terminology, either as sharing partners, characteristic of relations between consanguines, or as partners in ceremonial exchanges, characteristic of relations between affines. [1987:166]

It will be for the Onabasulu to sort out the problems in the event of implementation of the rules in the constitutions of the land groups. I have not seen the constitutions, and the Onabasulu I spoke with are unclear about them. If we look once more at the Manual of procedures for land group incorporation that was used by Chevron, we find the following in section II, b. 'qualifications for membership' in the section on the preparation of a constitution. (I quote at length, as it is also a particularly interesting example of the contradictions inherent in making 'law' of 'custom.')

> This is the most sensitive matter in incorporating land groups, where the greatest care needs to be taken. If the membership rules are not sufficiently clear, the likelihood of disputes is increased when money is being distributed to group members, etc. *It must be possible, therefore, to tell from a land group's constitution whether a particular person is a member of the group or not.* [Fingleton n.d., [1993]:6-7, emphasis mine]

This is, of course, a rather 'non-Onabasulu' requirement, for this was always problematic for them, partly in for reasons outlined in general terms in Lévi-Strauss' observation. Also, dispute is part of social organisation, especially in the process of *mosomu* segmentation and fission. The Manual, after pointing out that this is where 'the preliminary work on genealogies and social organisation in the area' is important, continues:

> The important requirement is to set out in the groups constitution- (i) the qualifications for membership; and (ii) any disqualifications from membership.
>
> These must be set out in a way which is both *sufficiently flexible to allow for the usual flexibility on such matters under custom, but also sufficiently precise to allow any dispute over a person's membership to be decided.* [Fingleton n.d., [1993]:7]

My 'authority' on such matters was clearly based on an assumption that I might be useful. Ole, a 'clan' of interesting status, is involved in a dispute over an oil well involving some Huli and (probably quite incorrectly) I was seen as a useful resource. People knew that I worked in Fasu in 1992 on the incorporation of land groups.

There is the problem that Ole is in a very real sense no longer a group. Hence the story of Wafoale, the last man in the line Ole. The territory of what was Ole is used by members of Hanoro, Sabiasulu and Sibisi lineages, acting in concert. The last man of Ole – who had the misfortune of having a tail, making him an unattractive prospect as a husband as well as giving him sitting difficulties – bestowed the land to the father's father of Wabue. Wabue, the last male of his segment of Sabiasulu, as he had only female children, bestows the land's custodianship to the sons of his father's sister's daughter's son.

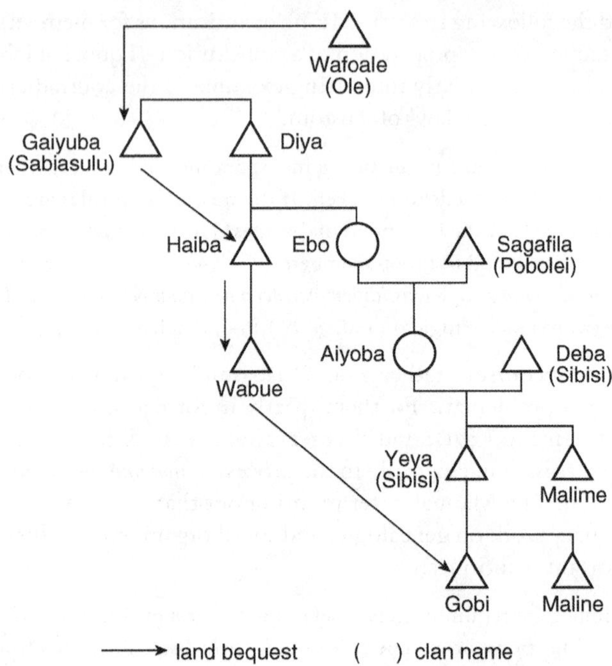

Figure 3: Relations in the story of Wafoale, the last man in the line Ole.

These are appropriate as they classificatory sons to Wabue. In Onabasulu kin terminology, both matri- and patrilateral female cross cousins are termed 'mother', so Yeya's (and Malime's) mother Aiyoba [another irony? a namesake of the old woman captured by Hides] is also 'mother' to Wabue. Yeya's sons are Wabue's [brother's] sons as well. And while they are away at school in Mendi, who better to take a place in the management structure of the 'clan' than Malime and his son Anton?

The processes are all appropriate in Onabasulu, and fit nicely with those social processes I saw as *implicit* in the 1970s. The irony is that they become explicit in contestation of control of an incorporated land group that is a 'dead' clan! (See fig. 3.)

The 'clan' is not an exclusive descent group, or any other kind of exclusive group in Onabasulu: conceptually people can clearly belong to a number of them. Similarly, the clans (lineages) are not permanent entities – they are historically contingent and dependent on the activities of men and the connectivity of women. Men act in history by intentionally creating, splitting and obliterating and now resurrecting these small groups. These activities, of course, take place in an arena of contest. Yeya is Wabue's brother. But not his only brother. Sharing a mother can arguably make them closer than 'clan brothers' of two or three degrees of agnatic collaterality, but the key word here is 'arguably'.

In the midst of this, then, what is a 'clan' – that landowner group the 'existence' of which, as an entity, is so beloved by many associated with Chevron and the Papua New Guinea Government? The Onabasulu know. It is a company (*kompani*), and as such implies business (*bisnis*). It has a chairman and a committee, and can be a component of larger companies that are formed by shareholders to distribute the monies and create business from the funds obtained from their resources (*resosis*). The Onabasulu see clans as companies whose business is *bisnis* and they asked me if I knew much about the development of management structures (yet another irony that would not be lost on the dean of arts and vice chancellor of my university). They may see them more clearly than a number of anthropologists, lawyers and company and government officials. The reality of the clan in Onabasulu is in a very true sense a 'legal fiction', and in this sense it merges with Wagner's discussion of anthropology's kin groups. 'Are there social groups in the New Guinea Highlands?' Wagner asked (1974). Maybe not, but they are incorporated under law now.

And there is a rub. The processes of action, contestation and development of these incorporated clans are precisely and necessarily those of *mosomu*. In 1969, Agibe Fua related to me his first encounter with money. He asked who made it, and why it could be exchanged for everything. He said it could change things radically, because of this quality. Although change things it did, it did not do so in the ways anticipated by me or by Agibe. What of clans? They are *kompani* and *mosomu*. Their usage on the peripheries of global capital has already given a reality to incorporated groups, that are and yet are not *mosomu* based in law, and a reflexivity to kin practices that are significantly different from those envisaged in the incorporation process.

Conclusions

Above I have explored the implications of three stories that were told to me in 1996. They were all told in a context of acute awareness of oil as a resource and its local politics, as people were following the unfolding of a small crisis at Kutubu, with a landowner company requesting more equity in the Chevron run project, and in the ensuing dispute, Chevron evacuating their personnel.

They represent the perception of a feature of the landscape modified to include oil and gas potentials and to define the Onabasulu. And in the process of this, ethnic relations in the area are reconfigured. The snake provides oil and gas, marks a border and provides connection for these things to long term Onabasulu beliefs.

The story of Duduma takes the Onabasulu definition further. The seventeen true Onabasulu clans are defined in both myth and landowner group certificates. There is a purposeful reality of Onabasulu being which was, in the recent past, simply not an explicit part of people's lives.

This entivity of ethnic category is matched by the reified clan – organised by the processes of kinship, yet having much in the nature of companies or businesses: part of the fluid non-totalising processes of Onabasulu social life, and yet with the firm reality of a certificate, a committee and dispute management procedures – straight out of Weber for Beginners.

Acknowledgments

This chapter is based on research carried out with the aid of a Doctoral Research Grant from the National Institute of Health (USA), and travel grants from the University of Papua New Guinea and Charles Sturt University. Some of the research was carried out while a visiting fellow in Anthropology in the Research School of Pacific and Asian Studies of the Australian National University. The final version of this chapter was prepared while I was a visiting research scholar at the National Museum of Ethnology, Osaka. I thank those institutions for their support. Earlier versions of this chapter have been presented to a conference at James Cook University in October 1996 and at Sydney University's anthropology seminar series in April 1997. I thank the participants in those readings and at the Myths to Minerals Conference for their comments. Especially helpful were comments by Steve Feld, Neil Maclean, Michael Nihill, Mike O'Hanlon, James Weiner and Kerry Zubrinich. Comments by Jadran Mimica were extremely useful and clarifying for me and Aletta Biersack carefully read and commented very helpfully on this chapter. Of course, all responsibility for the chapter and its shortcomings is mine, although its strengths owe much to these people.

Notes

1. The third edition of the *Shorter Oxford English Dictionary On Historical Principles* has the following entry for *Entify*: '*v. rare* 1882 To make into an entity, to attribute objective existence to. Hence *Entification*.'
2. As is also the case in 'reification,' entification specifically omits attention to how the entified object is constituted.
3. See Schutz 1962 and 1967 (especially parts I and III) as well as the discussion of Schutz in Thomason 1982, (especially chapters 3 and 4) for material on this perspective.
4. As well as in the more general conditions in Papua New Guinea sometimes called 'modernity.'
5. Caution, of course, is required in speculative matters about the direction of further change. Marshall Sahlins is fond of citing Durkheim's dictum that 'The science of the future has no subject matter.' What can be said, though, is that the discourse about and functions of ethnic and social categories in Onabasulu have changed markedly.
6. Laurence Goldman, after a visit to Walagu provided additional material and place names used this story. This is a corrected version of the story that appeared in Ernst 1999:91. Goldman goes on to suggest that "...what you have here is the use of Fasu (Auwaga=Folola) and Huli (Haliago = Huli term for Sisa) names which makes the Onabasulu rhetoric even more interesting because they adopt and utilise the familiar terms of discourse of their 'opponents'"(Goldman, pers. comm. 26 January 2000). It could also be a form of incorporation.
7. One thing that is interesting when distinctiveness is being carved from complex resemblances is who can go where and know what. The Bedamuni interact with the Etoro and fight with the Kaluli and other 'Bosavis'. They do nothing in relation to the Onabasulu. The Onabasulu had a story, told me in 1973 when I was trying to get some people to walk over to Bedamuni territory with me, that if Onabasulu cross the river that separates the Etoro from the Bedamuni territory, their legs are cut off at the height of the water. Etoro and Bedamuni can cross with impunity. But then, no one really knows if this was true. On my most recent visit, I was told that on Kebi territory, there is a site near where Duduma was killed where the water falls into the ground, emerging in Bedamuni territory, and forming a road for exchange. (There are rocks in the water here

as well, and some Kebi men can read aspects of the state of the ground by looking at the trailing weeds on the rock and their patterns in the flowing water.)

8. Onobasulu is the spelling used by the people themselves. Dondorp and Uringa state "The language name is Onobasulu. However, the Ethnologue lists the language name as 'On**a**basulu' (three-letter code: ONA) with the alternate name of 'Onobasulu'. Onabasulu is also the name found in the literature. The people say it is 'Onobasulu' without any alternate pronunciations (1999:1)." The name was established by the time I entered the area. It was seldom heard in the early 1970s, and I have always heard it with an 'open o'.

9. Isao Hayashi notes (pers. comm. 1994) an analogous story from the closely related Bedamuni culture. In the Bedamuni case, the woman named Dunumuni is short rather than large. She was reported as being without genitalia. Her death occurred 'A long time ago when everyone lived together,' and the fragmentation of her body brought an end to that period, and was, as was the fragmentation of Duduma, cosmogonic.

7

THE POLITICS OF PETROLEUM EXTRACTION AND ROYALTY DISTRIBUTION AT LAKE KUTUBU

Bill F. Sagir

Crude oil in Kutubu, like the gold in Mount Kare and Porgera, has its myth of origin, which James Weiner retells (1994). I do not want to repeat the myth here because I do not have any ethnographic knowledge of the Foi and Fasu people. I rely on the works of Foi ethnographers Langlas and Weiner, and short studies of the Fasu by Mark Busse (1991) and Thomas Ernst (n.d.). I spent only two weeks in Kutubu but have maintained an interest in the oil project and its impacts on the Fasu and Foi people. What I will do here is describe the political relations between the Foi and Fasu that have emerged as a result of the oil project, and their relationships with the state. I will also describe some ways in which my work on a consultancy arising from that project placed me in a difficult position, as a Papua New Guinean working within the framework of an Australian-derived land regime which vests all underground mineral rights in the state.

From clans to incorporated land groups

When I went to Kutubu in November 1993, the process of land group incorporation was in its final stages in Fasu, and almost as far along in the Foi area. (For the locations of these areas, see Map 7). The incorporated land group, or ILG, was the preferred vehicle for channelling petroleum revenues to indigenous landowners at that time. The way people seemed to equate the clans with ILGs fascinated me. They spoke as if the 'clan', a traditional mode of social organisation, and the ILG were one and the same thing. At first I could not understand this. I understood ILGs to be legally constituted business entities, which, in the case of Kutubu, were formed to function as channels through which Chevron could pay royalties and other monetary benefits. The incorporation of traditional social groups as business entities is provided for by the *Land Groups Incorporation Act, 1974*, the purposes of which 'are to encourage:

(a) greater participation by local people in the national economy by the use of the land; and
(b) better use of such land; and
(c) greater certainty of title; and
(d) the better and more effectual settlement of certain disputes,

by:

(e) the legal recognition of the corporate status of certain customary and similar groups, and the conferring on them, as corporations, of power to acquire, hold dispose of and manage land, and of ancillary powers; and
(f) the encouragement of the self-resolution of disputes within such groups.'

The ILG has a constitution which is recognised by the state. The *Land Group Incorporation Act* requires that at the time a clan or other customary group submits its application to be incorporated, it has to include names of all its members. When the clan group is incorporated, it is issued a certificate of incorporation. This seems to me to impose an overly rigid boundary on traditionally flexible groups. Such flexibility is shown by acceptance of migrants from other places by clans in Fasu and Foi areas. In Foi, Langlas and Weiner (1988) report that local clans commonly and easily accept immigrants. The hosts give these migrants rights to clan resources and political and ceremonial support and, if large, these immigrant groups may retain their separate identity permanently (Langlas and Weiner, 1988:78). The same seems to be true among the Fasu. Adoption is also common among the Foi and Fasu. Foi men 'care for' their own children in much the same way as they 'care for' any adopted children and I would assume that this would also be true of the Fasu. The rationale of ILGs seems to go against these aspects of Foi and Fasu social organisation. Can members of one ILG leave it and be 'adopted' or accepted into another ILG? Perhaps not, because: (i) the members of each ILG are already identified and the list of members submitted with the application for incorporation; and (ii) being a member of an ILG means obtaining monetary benefits from Chevron. The fewer members an ILG has, the more money each member gets. If new members are recruited it would mean that each member would get less money.

The rigidity imposed by the ILG is perhaps best exemplified in the case of Hami Yawari, a leader of the Foi Association. During the conflict between Fasu and Foi people over royalty distribution, Hami Yawari complained that part of his clan (*Yamagene*) lives in Mano village in the Fasu area where it is known as *Nigima*. Prior to the formation of ILGs, when Chevron paid compensation for damages done to Hami's clan land during drilling at Hedinia 1 and 2 (in Fasu territory), he collected the money and distributed it equally between members of his clan in both the Fasu and Foi areas. When the process of land group incorporation started, the part of his clan

Map 7: The Lake Kutubu area and environs

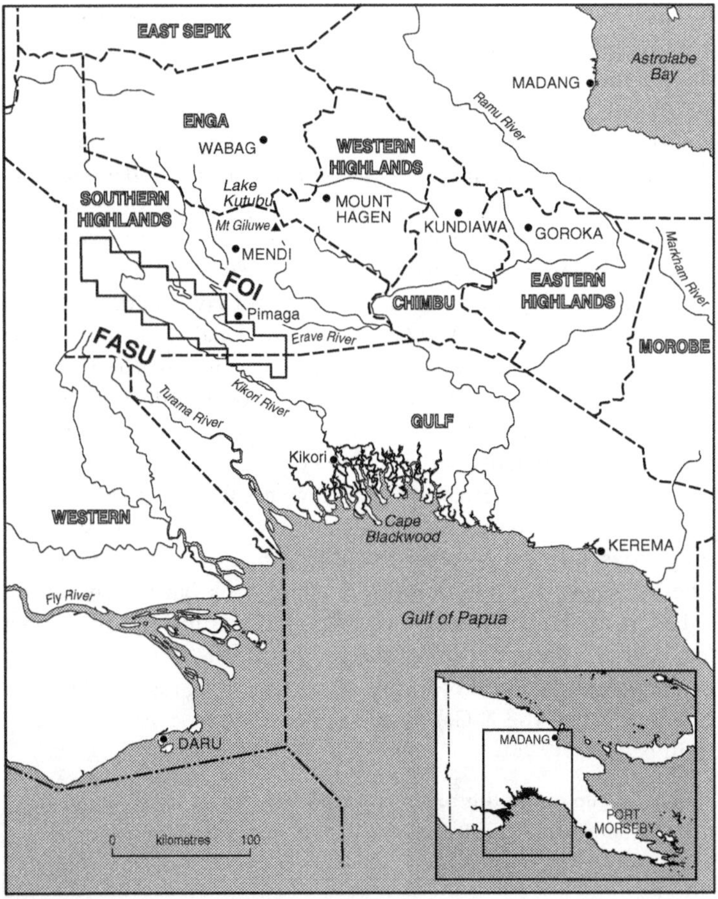

living in Fasu territory was incorporated as a distinct ILG, separate from the other part living in the Foi area, thus being entitled to more royalties than if they had amalgamated with the Foi part of the clan.

If the rationale of ILGs goes against the traditional organisational principles of Foi and Fasu people, why persist with them? The answer seems to be that it is easier for Chevron and government officials to deal with ILGs than with traditional social groups like clans. Through the ILGs, some form of bureaucratic control is imposed on these societies which had had little to do with the state and its bureaucratic machinery in the past. As Filer (1993:7) states 'The incorporation of the local community through the adoption or imposition of bureaucratic management methods in various spheres of activity' is one general process in the Papua New Guinea resource development scenario to which social impacts of a project can be attributed. Incorporation of clans as land groups also extends state control over people who had previously had very little contact with the state.[1] Land group incorporation puts the people under state laws (the *Land Groups Incorporation Act*, for one) and the state imposes taxes on incomes the people receive, as well as deducting management fees from royalties that the people receive.

Incorporated land groups are also easy for lawyers to deal with (cf. Ernst 1996, this vol.). ILGs are after all, legal constructs. In the court case taken out by the Foi landowners against the Fasu and the state, for example, the Foi wanted the court to declare that the national government decision relating to the distribution of royalties and other benefits between the Foi and Fasu landowners was unconstitutional. Against this, the lawyers for the state and Fasu landowners argued that under the Papua New Guinea constitution, only citizens are empowered to enforce their constitutional rights. The Yamagene Land Group (Inc.), which took out the lawsuit on behalf of the Foi ILGs was not a citizen. It was merely 'an abstract legal entity'. There is also the danger that Ernst (1993:6) mentions in the original version of his report to Chevron, that is, that such groups (that is, the ILGs) should not be seen to exist in the long term because:

> Local groupings in this region not only have upper limits of populations (which are exceeded only artificially and on paper in designated census centres), [but] as groups, they are structurally unstable. This is,

in part, because internal disputes can be resolved by splitting. [Ernst 1993:6]

The politics of royalty distribution

Under Section 118 of the *Petroleum Act Chapter No. 198*, the holder of a petroleum development license, in this case Chevron and its joint-venture partners, pays to the state 1.25 per cent of well-head value of all petroleum produced within the license area as royalty payments. Under the old Organic Law on Provincial Governments (which was repealed by Parliament in 1995 and replaced by the new Organic Law on Provincial and Local Level Governments) the state then pays 100 per cent of the royalties to the Provincial Government in the form of Derivation Grants. The Provincial Government in turn gives at least 20 per cent to the landowners.[2] In the case of the Kutubu petroleum project, on December 10 1990 the state signed one Memorandum of Understanding (MOU) with the Southern Highlands Provincial Government (SHPG) and another with the Kutubu landowners – which then enabled the granting of a Petroleum Development License (PDL2) to Chevron and its joint-venture partners.

Clause 6.1 of both agreements states that all royalties from the Kutubu project will be paid to the SHPG. The SHPG will then pay at least 20 per cent to the Kutubu landowners. From December 1990 many meetings were held in Kutubu villages, Mendi and Port Moresby to forge an agreement regarding, among other things,[3] the exact division of royalties between the SHPG and Kutubu landowners, and how the landowners' share of the royalties would be distributed between the Fasu and Foi landowners.

There were more intensive meetings immediately before the first oil shipment in June 1992. The SHPG offered 30 per cent royalties to the Kutubu landowners but the landowners did not want to accept this offer initially. When they eventually accepted the 30 per cent offer they demanded that their share of the royalties be paid directly to their ILGs by Chevron because they did not trust their provincial government. Government officials and people in the mining and petroleum industries saw this 30 per cent offer as a very generous one. SHPG seemed to be more generous to the landowners than other provincial governments. Enga Provincial Government, for example, gave 23 per cent of royalties to Porgera landowners and in the other mining projects, landowners have been given only 20 per cent.

After the suspension of the SHPG in October 1992, the Minister for Mining and Petroleum and the Southern Highlands Provincial Administrator held more meetings with the Foi and Fasu landowners. Things got complicated when the Fasu landowners began demanding 100 per cent of the royalties (with nothing for the SHPG and the Foi landowners). In June 1993, while the SHPG was under suspension, the National Executive Council (NEC) made a decision (Decision 101/93) that of the 30 per cent landowners' share of the royalties, 90 per cent would go to the Fasu landowners and 10 per cent to the Foi landowners. This was in recognition of the fact that 90 per cent of the land covered by the petroleum development license area (PDL2) was owned by Fasu people and that all the actual petroleum pools were in Fasu territory.

The Foi landowners did not want to accept this decision. They argued that some Foi groups were very closely related to Fasu people and had land, or rights to use land, in Fasu territory and that to be given only 10 per cent of royalties did not reflect this reality and was unfair. But this conflict between Foi and Fasu landowners goes back to 1989 when Chevron field officers began land investigations to determine ownership of land in the proposed PDL area. These were conflicts over the 'control and ownership of oil-bearing ground' (Weiner, 1994:47). With the NEC decision, conflicts between the two groups became conflicts over distribution of royalty from the sale of crude oil.

Despite the NEC decision in June 1993 regarding the distribution of royalties between Foi and Fasu landowners, royalty payments to the ILGs were held up because of delays in the certification of ILGs and also because of arguments between the Foi and Fasu landowners, and between Fasu landowners and the SHPG. Fasu ILGs were certified by early November 1993, and arranged a meeting with the Department of Mining and Petroleum in late November 1993 to discuss royalty payments. Mining and Petroleum officials did not turn up for the meeting. This resulted in Fasu landowners blocking Moro airstrip and stopping Chevron workers from going out into the oilfields.[4] Fasu landowners and their leader Sosoro Hewago demanded they be paid their share of the royalties within one week.

Senior Mining and Petroleum officials met with Sosoro and executives of the Namo'aporo Association (the Fasu landowners association) and promised them that Fasu ILGs would receive their royalties within three days. Mining and Petroleum officials, knowing that the government procedure for payments would take much longer than three days, approached Chevron and asked the company to make the payments. The Fasu ILGs did then receive their royalty payments within three days. The Foi did not want to accept their share because it was 'peanuts'. No Foi landowner was present at Moro at the time when Sosoro arrived with DMP and Chevron officials to present the cheques to the landowners. When their share was taken to Pimaga early the next morning, they rejected it. They demanded that their share be increased from the current 10 per cent.

A few days later, Foi landowners blocked Moro airstrip and the Kutubu access road. Police had to be called in to clear the airstrip and remove roadblocks. Leaders of the Foi landowners were arrested as well. In early 1995, Foi landowners took the state and Fasu landowners to court seeking remedies for what they perceived to be gross injustice in the way royalties were distributed between Fasu and Foi landowners. Sororo called a meeting with some Foi landowners and tried to persuade them to stop the court action. He told them that Foi people actually own only 8 per cent of the land within the PDL but Fasu people, out of courtesy and for the sake of peace and harmony, gave them an additional 2 per cent to increase their share to 10 per cent. If the Foi persisted with the court case, the Fasu would take back the 2 per cent and reduce the Foi share to 8 per cent.[5] The Fasu would also withhold other benefits that they are supposed to share with the Foi. The Foi leaders, however, persisted with the court case. In my view, the distribution of royalties between the two groups really is grossly unjust. Figures in Table 1 will shed light on this.

Table 1: Royalty payments (in kina) to Fasu and Foi landowners, June 1992 to December 1994

Year	Fasu share of royalties (90%)			Foi share of royalties (10%)		
	Fasu total	Fasu ILGs (60%)	Fasu future generation (40%)	Foi total	Foi ILGs (67.67%)	Foi future generation (33.33%)
1992	883 899.90	530 339.94	353 559.96	108 056.22	72 037.48	36 018.74
1993	2 327 955.51	1 396 773.31	931 182.20	284 591.14	189 727.42	94 863.72
1994	1 939 884.83	1 163 930.90	775 953.93	237 149.73	158 099.82	79 049.91
Total	**5 151 740.24**	**3 091 044.15**	**2 060 696.09**	**629 797.09**	**419 864.72**	**209 932.37**

From June to December 1992, Fasu landowners received K883,899.90 compared to the Foi's K108,056.22. In 1993, Fasu received K2,327,955.51 while Foi received only K284,591.14. In 1994, Fasu received K1,939,884.83 compared to the Foi's K237,149.73. From June 1992 to the end of December 1994, Fasu landowners had received a massive K5,151,740.24 compared to the mere K629,797.09 received by the Foi in this same period. The decision to divide royalty payments in the 90 per cent : 10 per cent formula clearly did not take account of the fact that some Foi groups have rights to land in Fasu territory and that the two groups have a long history of intertribal marriages. Foi leaders like Hami Yawari and Sese Vege state in their affidavits that in '*taim bipo*' Foi and Fasu people shared things together, that they had shared land and bride price. And a number of Foi clans had actually migrated from Fasu. Government anthropologist, F.E. Williams found that in 1938, seven of the eleven clans in the Lake Foi village of Yo'obo were of Fasu origin (1977:208). In addition, Weiner found that in the Foi village of Hegeso, two clans were of Fasu origin and migrated to Hegeso via Yo'obo in the Lake Foi area (Weiner, 1988:22). In early December 1993 Foi people told me and my colleagues during village meetings that in the past whenever Foi people killed pigs, they would share the pigs equally between Foi and Fasu, and that the Fasu did likewise when they killed pigs. Now, with royalty distribution, they thought this same formula should be followed; that is, Foi and Fasu should share royalties 50:50 between them. One interesting thing that was not mentioned during the conflict between Fasu and Foi over royalty distribution was the mythical source of oil in Kutubu. According to the myth retold by Weiner (1994), the source of the oil was a Foi woman from Baru who was accused of adultery and struck with a cassowary bone. She then fled and eventually ended up in Fasu territory, which is where the oil is now.

The Foi landowners eventually lost the court case on some legal technical matters rather than on substantial matters such as whether or not the formula for distributing royalties between the two groups is just.

Questions of resource ownership

In late November 1993, a colleague and I, accompanied by two Chevron Community Development staff, had a meeting with some men in Hedinia. After the meeting we proceeded to interview some of them. I spoke with a young man who put a question to me. He told me that his clan owned land where the Hedinia oilfield was located. He asked why it was that the state (or as he said, the *gavman*: ordinary village people do not talk about 'the state' as such) had come in between and claimed oil from their land. I tried to explain the *Petroleum Act* to him but he said he had heard all about that act, as it had been explained so many times to them by government and Chevron officials. He still had trouble trying to separate the oil in principle from the land which was theirs by custom. The land is ours, he said, and oil found in our land should be ours too.

I must admit that as a Papua New Guinean, I also have trouble trying to separate minerals and petroleum and natural gas, which according to Papua New Guinea laws, are the property of the state, from the land, most of which is under customary ownership. I was to learn later that one of the reasons why the Fasu people had wanted 100 per cent of royalties was because of the fact that all the actual oil pools were in Fasu land and that they considered the oil to be theirs. To be given 100 per cent royalties would be compensation for the state taking away ownership of the oil from them.

The Fasu leaders have also expressed anger over the fact that prior to the discovery and extraction of crude oil in the area, the state was an unknown entity. Governments since Independence had paid very little attention to the area. The only social services were provided by the Asia Pacific Christian Mission (which later became the Evangelical Church of Papua). Now with the discovery of oil, the state suddenly appeared and claimed ownership of the oil in their land. The Fasu people seem to me to have become satisfied with the royalties and other benefits they are getting and have not challenged the state in court over ownership of the oil. Landowners in the nearby Gobe oil project, however, have taken court action against the state over the issue of ownership of oil. (This was a case the landowners lost).

Conclusion

Landowners throughout Papua New Guinea are now beginning to question the state's claim to ownership of resources (minerals, petroleum, helium) that are found in the land (ground) owned by customary landowners. Wherever resources are found in the ground anywhere in Papua New Guinea, these automatically become the property of the state. Section 5 of the *Mining Act* 1992 states that 'All gold and minerals in or on any land in the country are the property of the State' and Section 5 of the *Petroleum Act* 1992 states that '... all petroleum and helium at or below the surface of any land is and shall be deemed at all times to have been the property of the State.' Most Papua New Guineans are like the young Fasu man who questioned me in 1993. They cannot separate the land from what is beneath its surface. For them, land includes everything that is in and on the land. Minerals and petroleum and helium are part of the land in which they are situated. Another Papua New Guinea law, the *Land Disputes Settlement Act* defines land exactly in the way that ordinary Papua New Guineans understand it: 'Land means customary land and includes (a) reef or bank and (b) house or structural building on land or over water, (c) things growing on land or in water, over land and minerals on or under land, and (d) interest in land.' If land includes minerals on or under land, as defined by this act, then surely customary landowners own the minerals and petroleum found under their land. It surprises me that the state can use other laws to claim ownership of these resources.

Since 1991, a number of Papua New Guineans have taken the state to court challenging the state's claim to ownership of resources found under land owned by customary landowners. In all these cases, one of the main arguments put forward by lawyers for the state has been that Papua New Guineans do not have a history of making use of resources such as oil and minerals found under the ground. Those who have challenged the state's claim have themselves been challenged by state lawyers to show 'personal or proprietary rights' to the minerals and oil found below their customary land. In the first of these cases (*Donigi v The State* [1991] PNGLR 276), the judge made extensive use of Thomas Harding's piece on land tenure in the *Encyclopedia of Papua New Guinea*, volume 2, to show that Papua New Guineans traditionally did not make use of oil, gold, copper and other minerals found under the ground. They did make use of clay for pottery and obsidian for tools, but not gold, copper and oil. As such, they cannot claim ownership over these resources.

Against this, there has been the argument that there has never been a traditional state of Papua New Guinea that made use of these resources in traditional times which would enable it to claim ownership over these resources. The modern state of Papua New Guinea is only making use of laws it borrowed from England and Australia to impose state ownership of these resources. The legal regime for mining and petroleum industries in Papua New Guinea has been borrowed wholesale from Australia, as Wolfers (1992:253) notes:

> Very few experts seem to be aware that the legal regime for mining in Papua New Guinea does not embody unvarying practice in advanced industrial countries, let alone that much of it is essentially (and peculiarly) Australian ... Alternatives have rarely been discussed. Even the (legally, narrower) suggestion by Papua New Guinean lawyer Peter Donigi that the legal vesting of minerals in the state might be unconstitutional has not been taken up. The prevalence – and dominance – of Australian influences, intellectual and material, in Papua New Guinea mean that discussion, even among Papua New Guineans critical of the *status quo*, has rarely extended to systematic questioning of Australian precedents and models, let alone to the proposal of radical alternatives.

So far, all the court cases challenging the state's ownership of petroleum and minerals found below customary land have been unsuccessful. Some Papua New Guineans hope that one day a Mabo-like decision will be made in a Papua New Guinea court that will give Papua New Guineans ownership of resources such as oil and minerals found in their customary land. The Mabo ruling was an Australian High Court decision regarding Australian Aboriginal land rights (see Merlan, this vol.). In Papua New Guinea most land is still owned by customary landowners. But they do not own the oil and minerals that may be found in their land. These automatically become the property of the state. In Papua New Guinea, we still have our land. We have not been dispossessed. But we do not have rights to minerals, petroleum and natural gas found in our land. As Donigi has put it:

> Well, we still have our land but our rights to gold and other minerals have been taken away by virtue of Section [5] of the Mining Act ... to be given to foreign corporations. [Donigi, 1988:25]

Notes

1. Prior to the discovery of crude oil in Kutubu and its eventual extraction, Kutubu people had had very little to with state and with 'gavman'. The Kutubu area was on the periphery of any form of development that the state initiated for the Southern Highlands Province. The Southern

Highlands Integrated Rural Development Programme extended to some Foi villages but did not go as far as the Fasu area. Some Foi villages planted coffee but transport was a big problem at that time.
2. Under the new Organic Law on Provincial and Local Level Governments, 100 per cent of the royalty is paid to the landowners. The provincial government misses out. This however, applies only to projects negotiated and developed after the passing of the new law in 1995.
3. The 'other things' include: determining how much of the Special Support Grant funding provided to the Province would be made available in the Kutubu area; the setting up of a Development Authority to allow the landowners some input into Special Support Grant infrastructure Programme; and the distribution of the 7 per cent project equity share in the Kutubu Project which is committed by the state to the landowners in the MOU signed on December 10, 1990.
4. I was in Kutubu at the time of this landowner protest. My colleague and I accompanied Chevron community development staff to a Foi village. We were stopped on our way back to Moro. The vehicle was taken from us and we were told to walk the rest of the way to Moro.
5. Official figures however, show that Fasu actually own only 81.2 per cent of the land in PDL2. Foi own 10 per cent while the remaining 8.2 per cent is owned by people in Mananda.

8

THE OLD AIRFORCE ROAD: HISTORY, MYTH AND MINING IN NORTH-EAST ARNHEM LAND

Ian Keen

In 1987 the Anthropology section of the Northern Land Council asked me to report to them on a dispute over the distribution of monies from the Nabalco bauxite mine at Gove Peninsula in Eastern Arnhem Land. The descendants of one patrifilial group ('clan') were accusing another group of the same moiety of wrongly claiming that group's land to itself, and of unfairly controlling the distribution of royalties from bauxite mining over that land. As a result I interviewed older people of several patrifilial groups over a period of several weeks, at Yirrkala, Galiwin'ku and Gapuwiyak. This chapter concerns an incident that took place during the course of that research, but which was peripheral to that particular inquiry.

The late leader (RM) of the Rirratjingu group told me he was concerned about establishing a line marking the end of Yirritja moiety country and the beginning of Dhuwa moiety land. At that time the mining machines were stripping the top 4 metres or so off what was said to be land of the Gumatj patrifilial group, Yirritja moiety, so the Gumatj Association received and distributed the royalties. But the mining was getting near the country of the Rirratjingu group of the Dhuwa moiety. There was a problem, however. Whereas divisions between one patrifilial group's country and another are fairly unambiguous in the ecologically rich zones of coastal swamps, rivers, estuaries and salt-flats, up in the sclerophyll woodland with its dry laterite soils the identity of land is indeterminate for many kilometres, until the watershed of a creek gives a firm indication of the group and moiety identity. RM wanted to establish as a boundary 'the old Airforce road' from the airstrip to Yirrkala, built during World War II when allied forces were stationed at Yirrkala (see Map 8). Bauxite mining had been going on for nearly twenty years, and the nearby mining town of Nhulunbuy was well established.

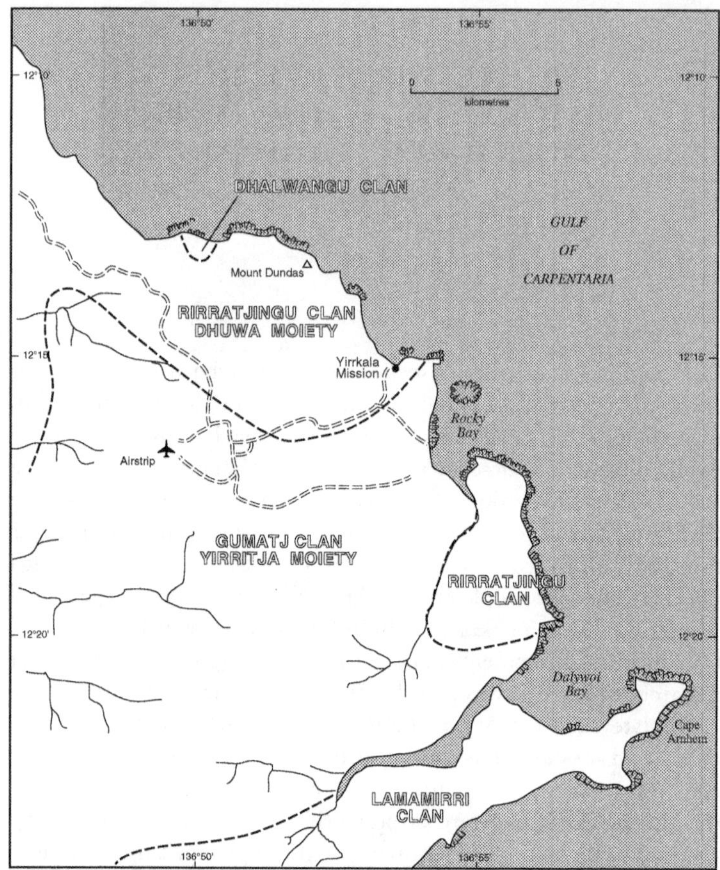

Map 8: The Old Airforce Road. Clan boundaries, airstrip and roads at Gove Peninsula, as shown in the Blackburn judgement in the Gove case (Blackburn 1971). The base map was taken from the 1:250 000 map using aerial photography of 1950.

The Gove bauxite mine

In 1965 the Federal Government granted a Special Mineral Lease over 140 square miles of land adjacent to Yirrkala mission, excised from the Arnhem Land Aboriginal Reserve, to mine bauxite. Smaller leases had previously been granted to to Comalco in 1958, and Pechiney in 1963. The Yolngu people living on the Yirrkala mission adjacent to the lease were neither consulted nor advised. With the assistance of the mission superintendent and a Federal MHR,

Yirrkala leaders lodged an objection in the Mining Warden's Court, and sent a petition in the form of a bark painting to Federal Parliament. A Parliamentary Select Committee was set up to look into the grievances of Yirrkala people, and made recommendations that Aboriginal rights over hunting areas and access to sacred sites be protected; to ensure effective consultation over future developments; to provide compensation for loss of reserve lands; and to ensure that Aboriginal people did not become fringe dwellers of the planned mining town of Nhulunbuy. The lack of effective consultation over the construction of an alumina plant, and an agreement to grant leases for forty-two years with a right to renew for a further forty-two years, prompted Yolngu people, helped by a missionary and a Federal member of parliament, to take legal action against Nabalco and the Commonwealth (Cousins and Nieuwenhuysen 1984). The case, *Millirrpum and others v. Nabalco Pty Ltd and the Commonwealth of Australia*, was heard in the Supreme Court of the Northern Territory in March 1970 before Mr Justice Blackburn. The plaintiffs sought declarations that they were entitled to occupation and enjoyment of lands free from interference, and that the relevant mining ordinance was void. They sought an injunction restraining Nabalco from interfering with the lands, and they sought damages. The case hinged on the question whether communal native title was recognised in common law. For various reasons Justice Blackburn found against the plaintiffs. The Gove case was a key event that stimulated the land rights movement in Australia, culminating in the enactment of the *Aboriginal Land Rights (Northern Territory) Act 1976*, under which the Arnhem Land Reserve became Aboriginal land. The mine continued, for the lease was on land that had been excised from the reserve (Cousins and Nieuwenhuysen 1984; and see Williams 1986).

The Commonwealth had agreed that the equivalent of royalties be paid to the Aboriginal community. Nabalco paid royalties for mining to the government, and the Commonwealth transferred an equivalent amount to the Aboriginal Benefit Trust Fund. The Dhanbul Community Association received and distributed funds for the Yirrkala community and related homeland centres. Half the funds, some $275,000 in 1981-82, were designated for use by the Association for community projects such as homeland centre development, and half were distributed to the two patrifilial groups, Rirratjingu and Gumatj, since land claimed by them was directly affected by the mining. Later, Gumatj people formed their own association to receive and distribute royalties when mining was on Yirritja moiety land. The Associations also received lease moneys from projects on Aboriginal land such as motels, and the mining company out-sourced contract employment to Yolngu Business Enterprises, which received a gross income of nearly $3 million by 1982 (Cousins and Nieuwenhuysen 1984:63).

These monies were the new land-based resources controlled and distributed on the basis of the patrifilial group and moiety identity of the land in question, that stretched existing modes of identifying and owning land and waters.

The old Airforce road

During my visit to Yirrkala, sitting outside his house at Yirrkala, RM explained to me:

> The old dirt road, the Airforce road, comes down to Yirrkala between Dhuraka and Gunyipinya, the two 'boundaries', Yirritja and Dhuwa.
>
> The airforce road opens up on this the Gunyipinya side, the Dhuwa moiety place. The other side is Yirritja – the hill where the mining is now going on. Father said to the family, 'When they take the mining on to the Gunyipinya side, the money has to be paid only to Rirratjingu, not Gumatj.
>
> The old road is the boundary here, as everybody knows – Gumatj and Rirratjingu all understood this.[1] At Gunyipinya the hill and bridge are all in Rirratjingu country ... The old people [*dilkurruwurru*] gave the word; we can't change their word.[2]

RM and I drove west to where the sealed Yirrkala road comes near the old airforce road. He pointed out Dhuwa moiety country on the left of the road, Yirritja land on the right. The strip of new mining, he said, is part of Gunyipinya, which belongs to Rirratjingu and bottom Djambarrpuyngu groups, connecting through the saltwater songs of Black-capped Tern to Djuwarr'nga, country of top Djambarrpuyngu Dhamarranydji group. The boundary comes from the river, and ends on the south of the bitumen road, RM explained. We drove west along the old road, then later returned east to Yirrkala, where we talked.

Back at Yirrkala, RM listed Dhuwa moiety places in the Nhulunbuy area, and said that what he had told me was an 'old history story from way back before mission came'. Leaders of several Dhuwa moiety patrifilial groups made an agreement that all the land be 'handed up' to Rirratjingu *ba:purru* to look after on their own. Many old men of one such group died out, so that their land, sacra and songs went to their Rirratjingu *gutharra* (women's/sister's daughter's children). RM went on to explain how the Rirratjingu Association would handle the money from business on Dhuwa land, and would share it with everyone.

I must confess to being sceptical. Certainly RM did legitimise the road as a boundary by saying it was agreed by the old people. But here was a mark of relatively recent human activity, whereas Yolngu talked of the identity of land and waters as ancestral givens. How likely was it, I wondered, that others would take seriously the proposal that the old airforce road should be taken as the boundary? The basis of my (unvoiced) scepticism was an unreflective and implicit distinction between, on the one hand actions and events in recent local 'history', and on the other, events of ancestral action, belonging in the domain of 'myth' and 'religion'. It is the basis of this scepticism that is the subject of this chapter, rather than the immediate politics of establishing a boundary. Perhaps in Yolngu ontology the old Airforce road is closer to the traces of ancestral *wangarr* in the land than the 'history/myth' or 'secular/religious' distinctions suggest. Some recent writing on Aboriginal history and myth (Sutton 1988; Rumsey 1994) suggests that at the very least these domains interweave in Aboriginal discourse. Before pursuing this line of inquiry, though, I want to explore those features of Yolngu land tenure that gave rise to the problem of demarcating a boundary in the first place.

Religious ecology in Arnhem Land

The 'estates' of patrifilial groups do not have boundaries in the sense of a perimeter that encloses and defines an area. The identity of country has as its foci places, often waters, interpreted as the traces of *wangarr* ancestral action. However, country is replete with markers of the end of land and waters of one group and the beginning of another. Such markers take the form of a change in gradient, hills and cliffs, streams, rivers and watersheds, changes in vegetation or soil, and at sea, changes in the colour of water (Williams 1986:82).

Both focal sites and boundary markers have the greatest density and clarity in areas rich in the resources salient to a hunting, gathering and fishing way of life, including technological resources such as stone and ochre. In the higher open Eucalypt forest and woodland away from the ecologically rich rivers, estuaries, swamps and plains, and stands of vine-forest, resources are scantier, and moiety and patri-group differentiation is less clear. The main resources of the Eucalypt forest, in which inland bands camped in the wet season, were macropods and honey, most easily obtained between December and May. But access in this season to the vegetable and fruit resources of the monsoon forest and fan deltas was essential, so that wet season camps had to be close to these resources. Fish, turtle eggs, most amphibians and reptiles, most birds, and file snakes are to be found in the richer zones of freshwater streams, rivers and swamps (Peterson 1973:183), and marine fish, reptiles and shellfish on the coast and in the mangroves.

Patrifilial group identity and 'holding' of land correlates with this distinction between resource-rich and resource-poor areas in two ways. One is the definiteness with which patri-group and moiety identity are assigned and marked ('definite' in both the presence and unambiguous character of markers, and the degree to which people agree). The second is the significance of the ancestral and other beings associated with place. The ecologically rich areas belong to major, 'heavy', ancestral beings (*wangarr*), associated with secret as well as public ceremonies, songs and sacred objects, while the poorer areas of sclerophyll forest belong to the 'Ghosts' (*mokuy*) and other figures of the public songs and dances, such as Murayana or Wandjuk (Stick Insect).

Distinct patterns of ownership mark the two genres of songs and ceremony. Each series of public (*garma*) songs of the forest is shared by several patrifilial groups of the same moiety. While each group possesses its own rather distinctive tune (albeit shared with other groups), and a particular selection of song and dance topics (each shared with several other groups), members of several groups with 'the same song' are able to cooperate in a single performance, either singing together, performing *seriatim*, or each incorporating some songs of another group within in its own series. Disputes among groups of the same moiety may arise not so much over the ownership of songs as the relation of songs to places. Elders of several groups may agree on which groups share songs pertaining to the area at issue, as well as other areas. But each may put his or her own group at the head of the list as those who truly owned the songs and the place, while the others merely 'sing' them.

The major *wangarr* ancestors, regarded as the creators of each group, its country, language and ceremonies, are associated with the resource-rich areas. Ownership of the associated sacra, while shared, is rather more clearly group-specific than the public songs. The most sacred, 'inside' *likanbuy* ('of the elbow') designs are painted on sacred objects and the body of initiands and the dead, or realised in simplified form as sand-sculptures. The *likanbuy* designs depict the results of ancestral action at a particular place, and differentiate most (but not all) group identities quite clearly, as do the *rangga* sacred objects and the related songs, dances and ceremonies. Each goes with one or more *likan* names that connote the identity of an ancestor, a place, and the group that bears them (and whose members now take those names as surnames). While some groups share *likan* names, just as some share very similar designs, most are distinct. The ownership of resource-rich areas is therefore harder to challenge by groups of the same moiety than the sclerophyll forest and woodland.

There are problems with boundaries, and these differ, depending on whether the abutting areas are of the same or opposite patrilineal moiety (Dhuwa or Yirritja). In several cases known to me where two countries of groups of the same moiety abut, and where there is no clear boundary, the country is said to be shared, 'company' land. In the case of adjoining land of groups of the opposite moiety, when discussing the extent of their estates one group is likely to push the extent of their land to a point where the identity of the country is indisputable, such as near a creek. The other group will push the extent of their land the other way. Where key resources are at stake, disputes among same-moiety groups are just as likely as between groups of opposite moieties.

What is to be done when new resources from relatively undifferentiated land have to be distributed? The bauxite at issue lay under woodland, just where patrifilial group ownership was least distinct. RM urged that the old Airforce road count as the boundary between Yirritja moiety and Dhuwa moiety land. Was he merely clutching at a straw, or did this mark in the landscape have more potential weight than might appear at first sight? On the face of it, the old Airforce road is an unlikely candidate as a group boundary because, in a 'Western' reading, it is the product of recent history. The legitimate and agreed markers of a group's connection to country are places with ancestral significance – traces of the presence, substance and activities of *wangarr* and other spirit beings whose activities are recounted in myth. But is this radical distinction between 'history' and 'myth' appropriate to the understanding of Yolngu ontology?

Myth and history

Recent discussions of Aboriginal narratives about the past retain the distinction between 'myth' and 'history', but seek to bridge the gap. Even in the West, the distinction is a relatively recent one. Writings on historiography show moves towards the idea of an objective and 'scientific' history relying less on received authority and more on archival evidence, took place gradually in the transition from the Medieval world, through the Renaissance to the Enlightenment (for example, Breisach 1983). The use of the word *myth* to denote a primarily fictitious narrative dates from the 1830s, according to the OED. In dictionary definitions of myth the sense of fiction predominates: a myth is a 'purely fictitious narrative usually involving supernatural persons, actions, or events, and embodying some popular idea concerning natural and historical phenomena' (SOED).

Anthropologists who have recently discussed this issue adhere to a distinction between myths and other kinds of narratives. The Berndts adopted the category 'verbal literature' and distinguished between myths and legends; the latter are largely about actual events in the past, although often distorted in transmission. By implication, myths are fictitious. Sacred myths are related to ritual, secular myths or folk tales are concerned with unreal events, universal themes such as jealousy, and spirits (Berndt and Berndt 1958; Hiatt 1975:1-2). In a later publication (Berndt and Berndt 1964) the Berndts revised this classification: stories are divided into sacred stories or myths, and ordinary 'stories' or oral literature. Ordinary stories are of two kinds, traditional and modern (see Hiatt 1975:2). Either way, the Berndts understood myths to be fictional narratives, at least some of which were associated with ritual or the sacred domain. Hiatt (1975:3) broadens the definition of myth to mean 'a traditional narrative which *in part* describes things that do not occur' (1975:3, emphasis in original). Hiatt discusses scholarly interpretations of myths as distorted 'history', reflecting actual events; as 'charters' for rites and social norms; as rationalisations of the wish-fulfilment content of dreams; and as ontology. It is clear that those scholars discussed by Hiatt who interpreted myths as reflecting real events did not go so far as to admit them as 'historical' accounts, but as 'echoes of a real past history' (Mathew 1899); as accounts that 'refer to a former time' and 'reveal' a state of organisation and customs (Spencer and Gillen 1899); or as 'phylogenetic records' of early states of organisation as postulated by Freud (Róheim 1925).

As Howard and Frances Morphy (1984) interpret Barthes:

> mythology is clearly incorporated within historical process and is part of the way in which history is understood. Myth is in part history, since it enters people's consciousness and affects their understanding of the world.

The Ngalakan narratives of their study were not recollections of the past

> but part of present understandings of the past that need bear no relation to what actually happened or was. History is important to our analysis not as a record of events but as a means of understanding how the relationship between the past and present has been continuously reconstructed and how the myths about the past fit the conditions of the present. [Morphy and Morphy 1984:462]

In their conclusion the Morphys find that Ngalakan 'historical myths' are 'biased in a number of ways' when compared with 'Roper River history' derived from other sources. Thus they retain the distinction between myth and history, with the clear implication that myth is largely fiction.

Sutton (1988) also retains the category of myth as distinct from other categories of narrative, and distinguishes stories in which the key figures are particular kin of the living, usually with remembered personal names, as distinct from myths in which key figures are generalised ancestral beings or agnatic forebears of group members as a whole. The personnel and events of myths are objectified, and relative to groups. While Dreamings too are people, mythology serves as an idiom, a 'legislative code in the third person' in which relations, histories, claims to land, and relations to totems are negotiated (Sutton 1988:253).

Having kept the distinction, Sutton goes on to show how such narrative genres may be mixed. Some stories about kin of the living are interwoven with Dreaming events. As Sutton points out, Aboriginal categories of narratives and places do not always distinguish kinds of stories along the same break as between 'history' and 'myth' in English. People of the north of South Australia refer to Dreaming ancestors as 'histories' in English, and to sites as 'history places'. Elsewhere the category 'story-place' is common. (RM's reference to 'old time history story' is similar to these usages.) Sutton understands site-related Aboriginal myths to be not merely invented pasts, but as in many cases a combination of invention and memory. Ancestral stories reflect events when, for example, travels of ancestral beings in stories closely follow the life histories of actual persons. In the desert, accounts of the experiences of living people merge in the telling with Dreamtime encounters; applying the term 'historical narratives' to such stories therefore has its drawbacks. Not only are such genres interwoven, but the approach to 'historical' and 'mythological' evidence is the same, Sutton argues. In both, the narrator constructs a version of events that may bear some relationship with events recorded in documentary sources, but not constrained by them. Sutton refers here to Aboriginal stories about Captain Cook's travels to places not recorded in his log (Rose 1984; Sutton 1988:256).[3]

More recently, Alan Rumsey (1994) has sought to draw out the implications of some 'characteristically Aboriginal forms of social memory' for attempts to rethink the dichotomies of myth/history and orality/literacy, which he sees as homologous. The relevant difference between orality and literacy has less to do with the difference between myth and history as modes of orientation, than with 'a potentially more open-ended range of what we can call "inscriptive practices", the relevant Aboriginal one being the use of features of the landscape as a medium for the production and reproduction of meaning'. This is a general orientation 'through which a good deal of what we call 'history' – the past action of known human beings – are also inscribed in and retrieved from the landscape' (1994:116).

Rumsey (1994:118) parts company from those, such as Stanner and Myers, who depict 'the dreaming' as an alternative to history, or as 'an erasure of the historical' (Myers 1986:69). To these authors 'the Dreaming' seems to be a quintessential example of what Turner (1988) refers to as a 'mythic mode of consciousness'. Drawing too great an asymmetry between the Dreamtime and the everyday as aspects of existence is problematic, Rumsey finds (1994:119). Reifying the Dreaming is compatible with identifying a whole people with mythic consciousness, a move that is both 'politically unfortunate' and contrary to his experience in the Kimberley. The Dreaming is rather a 'mode of orientation' and a way of experiencing the world (Rumsey 1994:120). He seeks to show that the characteristic kind of 'objectification' identified by Munn (1970) and others more or less exclusively with the Dreaming, 'figures in many other kinds [of narratives] as well, and in many other much more mundane aspects of Aboriginal people's social lives' (Munn 1970:121).

Pintubi people encode events in the Dreaming as well as 'the phenomenal world' in named places. Similarly, in a Ngarinyin narrative, events in the *larlay* ancestral period, the presence of a station manager, the performance of a divination, the death of an Aboriginal stockman, and conception, were all 'memoralized in the landscape' (Rumsey 1994:123; Myers 1986:68). Rumsey found that among Jawoyn people 'Dreamtime/mythical' and 'historical' ways of reading country make a similar use of the trope of punctuated movement through a fixed series of named places (Rumsey 1994:124). And near Fitzroy Crossing one man's exploits are memorialised in Winjana Gorge 'in just the same way as those of the primordial creator figures at many other places in the landscape' (Rumsey 1994:126). In sum, Rumsey avers that it is not a question of mythic versus historical consciousness, but the 'realisation of both' in a particular form of inscription in the places through which a person moves in the course of social life (1994:127). In this mode 'history' is not equivalent to the history characteristic of literary traditions, but is far more negotiable and subject to alternative interpretation than its literary counterparts, a point similar to Sutton's (Rumsey 1994:128).

The main basis of the distinction between myth and history made by these and other writers is that myths recount ancestral or other fictitious events, histories tell of events involving known persons. Personages in myths often bear a more generalised relation to groups than in narratives about named kin, and have to do with negotiated 'law' governing a number of domains (including land and ceremony). These modes are distinct in other ways as well, but may merge through the interweaving of ancestral and human events in the telling, similarities in their inscription of events and persons in places, the identification of persons and their histories with those of ancestors, and similarities in their relation to evidence by contrast with history in literate traditions.

Yolngu categories of narrative

Yolngu languages do not have equivalents of the English categories of 'myth' and 'history' – narratives about *wangarr* ancestors, and those concerning known persons are all *dha:wu*, 'stories'. While I do not want to minimise the differences between *wangarr* and ordinary people in Yolngu ontology, it may be helpful to do away with the myth/history distinction, and consider different kinds of stories in relation to Yolngu concepts of time. I want to suggest that to construe Yolngu cosmology as implying a dual division between 'the Dreaming' (which is not a Yolngu concept) and the phenomenal world of the present and recent past is not justifiable. It is more a question of relative temporal *distance*, what kinds of being figure in the narrative, and the relation of narrative to ritual.

Dha:wu is the generic term for 'narrative' – any account of actions and events from a piece of salacious gossip to the most formal ancestral story. Zorc (1986) defines *dha:wu* as 'story, word, news, information; message; meeting; speech'. The word can be qualified by adding the subject matter or kind of subject matter to which the narrative pertains, for example *dha:wu ngurru<u>n</u>anggal'wuy* (story about olden times), *dha:wu nga:thilinguwuy* (old story).

On the face of it, we might be inclined to follow traditional anthropological practice, and classify narratives about creator ancestors (*wangarr*) and other figures celebrated in rituals as 'myths'. There is some sense in this; a group's corpus of such narratives, associated with its complement of songs, ceremonies, designs and sacred objects, are elements of *rom* ('religious law'). Someone who wants to outline the group's *rom* may well begin by telling the listener its key stories about *wangarr* ancestors and other figures in events of the distant past (*baman'*). Ancestral stories appear to occupy a distinctive domain in their relation to rituals. The world of 'myth' and of ancestors might be said to be discrete because of its intimate relation to that set-apart Yolngu domain of *ma<u>d</u>ayin* – of song and dance, painted and ground-sculpted design, specially designated places and sacred objects.

Exegesis of the meaning of songs used in public ceremonies brings known persons into proximity with distant ones. It is not just that singers can use a song about ancestors as an allegorical comment on current events and relations, but that the 'inside' meaning may close the temporal gap. For Ganalbingu singers, for example the 'inside meaning' about songs ostensibly describing ancestral Macassans and their actions was the activities of Christian Missionaries. Where the song described unloading rice from the prahu, the meaning was the unloading of flour from a lugger, and so on (Keen 1994:164). There are two possible interpretations (which may coexist) as to the implications for concepts of time – either the early missionaries are beginning to gain the status of *wangarr*, or the song and its exegesis are drawing parallels between relatively distant and relatively proximate events, and interweaving them.

Nevertheless, ancestral stories are stories *for* or *about* shared (whether secret or public) and stereotyped ritual forms. Those forms are objectifications not of known persons, living or dead, but special classes of beings. Even though this domain intrudes upon and interweaves with that of known persons and everyday events, it is nonetheless set apart. It is a domain that has to be conjured up, after all, in the Yolngu imaginary, through mimetic dance and song, the use of body paint, creating particular artefacts and places and the rest. There is some sense in thinking of the stories relating to this imagined and enacted domain, the beings and events they evoke, as inhabiting a separate domain as well. This is not to admit the straightforward applicability of Durkheim's (1915) division between 'sacred' and 'profane'. Modes of existence outside ancestral law (*rom*) are conceivable – the state of *wakinngu* ('wild', 'illegitimate') is a case in point; but the Yolngu everyday world of social interaction, social relationships, foraging, fighting and working, is governed by *rom*, even in a changing world. Furthermore, ritualised expression is ubiquitous – for greetings and partings, fighting and dispute resolution, as well as major life-cycle events.

There are other ways of categorising stories; some count as 'inside' or 'underneath' (*djinawa*, *djinaga*), while others are 'outside' (*warrangul*). While the contrast between 'inside' and 'outside' meanings is important to the religious domain (*ma<u>d</u>ayin*), where it denotes matters of more or less secret knowledge from more or less public ones, the contrast is not restricted to that domain. It can also indicate the contrast between an allegory and its referent in an everyday setting, for example. Some inside stories about ceremony and ancestors may be classed as 'heavy' (*ngonung*), implying severe restrictions on their dissemination.

Perhaps stories about *wangarr* ancestors are distinct from other kinds of story because ancestors are very distinct kinds of being who lived at a particular time, the 'Dreamtime'.

Ancestral stories

Wangarr are distinct in a number of ways. People insist that they were *people* (*yolngu*), but they had extraordinary powers, such as the ability to give birth on a grand scale, to leave comparatively massive traces of their presence and actions, in the form of creeks, hills, rocks and the like, and in their power. Most have a multiple identity in story, dance and song – as persons but also objects such as a hollow log, or a species such as a honeybee. Events can be placed relative to the active presence, long ago, of these beings. Transformations described in ancestral stories, such as a dog turning into a rock, or a person becoming a bird, are examples of 'becoming *wangarr*'.

Not only *wangarr* ancestors figure in ancestral stories, however, but also people with ordinary powers, who interacted with the *wangarr*. These include the ancestors who bore the group's 'surname', such as Garrawurra in the case of L̲iyagawumirr people, and men who stole the dilly bag and sacred objects from the Djang'kawu sisters. Other stories recount the doings of named 'ghosts' (*mokuy*) and other honey hunters, important in the public songs. In one L̲iyagawumirr story, often told to me, the person or persons who crossed the water from L̲iyagawumirr country to give songs and make waterholes for another group, are ambiguously the Djang'kawu ancestors or Garrawurra, an ancestor who can be placed at the root of a genealogical tree. Some genealogies place this ancestor only five generations back (or rather down at the 'roots', *l̲uku*) from the oldest living people. People conveyed the distance of ancestral events by saying, 'Our fathers, did not know them, nor our father's fathers, nor their fathers – long ago!'

Whereas the events recounted in ancestral stories occurred in far off times, not merely do the traces of ancestral activity and substance remain in the landscape, but many of the *wangarr* are still alive, usually deep under the water. Furthermore, people tell of their own or relatives' encounters with manifestations of *wangarr*, and especially a particular kind of *wangarr* called *malagatj*, dangerous serpents. In evidence presented to the Aboriginal Land Commissioner in 1982 Dhulumburrk explained what happened when he encountered an extraordinarily large snake at on the mainland near Milingimbi, a manifestation of the Lightning Snake. If one sees it approaching it is necessary to talk to it in its own dialect and say, 'Go back the way you came … Go back to the water into your waterhole …' If a person goes hunting the snake may come up to the person, who should give him mangrove worms (for example) to eat first. If not the snake will bind himself around and kill the person. In perhaps the 1940s a fisherman or crocodile hunter anchored near the mouth of the Woolen River. The snakes wound around the anchors and nearly pulled it down (Aboriginal Sacred Sites Protection Authority 1982:115-9). It seems that ancestral stories describing how a human-like *wangarr* became an animal or an object such as the moon, provide for the possibility of interpreting creatures and objects of everyday experience as manifestations of *wangarr.* So does the 'inside' location of *wangarr* such as Mangrove Log who moves in and out with the tide deep under the water, and the Lighting Snake who may snare a boat's anchor. (It is this kind of discourse that leads Rudder (1993) to divide Yolngu experience into everyday and a timeless 'inside reality'.)

Stories about *wangarr* ancestors, or manifestations of the *wangarr* may, therefore, recall events during the lifetime of living individuals. One such story explained the disappearance without trace of an infant on Howard Island in 1980. A *marrngitj* 'doctor' from the west of the region suggested that the child had been taken up into heaven by Jesus (assimilated with God to the category of *wangarr*). Fellowship worshippers at Howard Island later saw a bright light, in its midst the transfigured face of the girl, such appearances continuing for several weeks (Bos 1988:112-3).

Stories about ancestral journeys and relations among *wangarr* may relate to recent events in another way: they may be interpreted as encoding the exchange of sacred objects related to those *wangarr* between two patrifilial groups. Rudder interprets stories such as the one about Dingo catching the scent of Whale as 'oral histories of the establishment of ... relationships told in the guise of the travels of natural species in 'ancient times'' (Rudder 1993:65).

People talk about ancestral traces and human traces in the country in a similar frame. On a visit to look at traces of the Djang'kawu sisters at Ga:rriyakngur, country of the L̲iyagawumirr group, the late Ba:riya identified a group of paperbark trees with the Ma:tjarra string of patrifilial groups (*mala*) with Djang'kawu as ancestor. Three trees planted by the sisters stand near Djurrpum's hut, intended to be the beginning of an outstation. The Nga:rra ceremony ground used by Ba:riya's fathers and fathers' fathers was the very one from which men crept out to steal the dilly-bag and sacred objects from the Djang'kawu. Nearby, sticks standing in the ground marked the height of young men who camped there. An old warriors' track led from Ga:rriyakngur to the creek where the Djang'kawu landed on their return from Galiwin'ku (Keen 1978).

Nevertheless, Yolngu talk of ancestral actions and events as happening long, long ago (*baman'*), in 'far off times' (*barrku*, 'distant' in space or time). Warner (1958:10) understood *baman'* ('bamun') to refer to an ancestral period, much as Williams and Morphy gloss *wangarr* to refer to a period of time when the *wangarr* (in the sense of ancestral beings with great powers) were active (Williams 1986:28; Morphy 1984:17). If this is right, then 'myths' are stories about that ancestral period, the 'Dreamtime' of popular discourse about Aboriginal cosmologies.

Howard Morphy (1990) analyses what he sees as an intermediary category of Yolngu narratives that he calls 'myths of inheritance' as distinct from 'myths of creation'. Myths of inheritance 'are the means by which myth is adjusted to accord with political process'; people of a 'clan' project their ownership of a given set of lands and their associated sacra back into the ancestral past, so that it can be said to have originated then. However, the intermediary category of myth 'creates a disjunction between the ancestral past and the time when humans occupied the earth, layering the ancestral past so that some of it is more distal from human existence', although the demarcation of these 'layers' of time is not precise, and there is much interpenetration and ambiguity (Morphy 1990:314). Some Yolngu myths seem to accept a disjunction between the ancestral past and the human world, and resolve the problem by creating beings who 'have a foot in both worlds', and 'who occupy the space between the fading away of the creative ancestral beings below ground and the occupancy of the world above ground by humans'. Humans who were the founding ancestors of the clans also occupied this space (Morphy 1990:314-15).

While the distinction between myths of creation and myths of inheritance is a useful one in capturing aspects of the significance of narratives, a temporal 'layering' of events is not evident in the corpus of L̲iyagawumirr patrifilial group. In the story mentioned earlier, in part a 'myth of inheritance' in Morphy's terms, in part a 'myth of creation', the Djang'kawu sisters, or Garrawurra, travel across the strait from the mainland to Elcho Island to give Gun̲birridji people the slow *bil̲ma* songs for the Nga:rra ceremony and create fresh-water springs. The same teller on one occasion made the substitution: 'Djang'kawu, or maybe Garrawurra ...'. Thus there is no clear layering of time here. In a sense, genealogical time may be said to be 'layered', however, as when some L̲iyagawumirr people took Garrawurra, created by Djang'kawu, to be the apical ancestor five generations back from the senior living generation.

Some readings place events and their traces such as the old airforce road in an 'historical' time frame of more recent events. Others such as those of Swain and Rudder, following Stanner, see the Dreaming as 'everywhen', a timeless domain. Morphy (1990:314) adumbrates such a view. Although stories about encounters with ghosts are not uncommon, he suggests, few stories tell of encounters between living people and ancestral beings. The power of ancestral beings but not their physical presence intervenes in the world, except through the signs that represent them: 'The Dreaming was in the past and is in the present, but in the past it was here on earth and resulted in the creation of the landscape, whereas today it exists in a different dimension' (Morphy 1990:314).

To sum up, there is distinct genre of ancestral stories, relating ancestral events long ago. However, *wangarr* lie 'inside' the country, especially deep waters, and places are redolent with the traces, powers and transformed substance of *wangarr.* In explaining the significance of places, people talk of ancestral and recent human events in the same frame, and tell of encounters with manifestations of *wangarr* as well as ghosts of the dead. Ancestral stories may encode human events and relations, for example as 'myths of inheritance'. While ancestral events occurred 'long, long ago', ancestral stories do not consistently 'layer' events in time. Some commentators depict the ancestral domain as 'timeless' – both past and parallel with the present. John Rudder's (1993) analysis of Yolngu concepts of time may help resolve these somewhat contradictory interpretations.

Yolngu concepts of time

In a recent discussion of Aboriginal concepts of space and time, Tony Swain argues that 'Aborigines themselves do not, or at least once did not, understand their being in terms of time, but of place and space' (1993:2). Rejecting the view that Aborigines have a concept of time as 'cyclical', and following Ricoeur's suggestion that open-ended linearity of time requires the application of numbers, Swain suggests that 'it is best to state that Aborigines operate from an understanding of *rhythmed events*' (Swain 1993:19, emphasis in original). The concept of time implies that it is prior to events – an arena in which events occur. For Aboriginal people there was no fashioning of time, linear or cyclical, but rather 'a sophisticated patterning of events in accordance with their rhythms' (Swain 1993:20). The 'Dreamtime' is not a time at all, but rather 'abiding events', which 'do not change and for which history is irrelevant' (Swain 1993:22-3). Swain assures us that he is not suggesting that Aboriginal people experience a world devoid of past, present and future, but rather that 'they avoided allowing these to be worked into an ontology which conceded the sovereignty of time'. Abiding events are linked to rhythmed events though place; linear time was a 'fall' from place (Swain 1993:23, 27).

One can only agree with Swain that before the adoption of clocks and watches, Yolngu concepts of time did not incorporate an independent, numerical, linear construct of 'time'. However, such a construct does not necessarily imply time independent of events, but rather takes certain events – the rotation of the earth and the annual cycle of seasons – as reference points in an imagined temporal space against which to position other events. 'Abiding events' sounds like an oxymoron, unless the event is continuous, such as the expansion of the universe, in which case the word 'process' seems applicable. (The word 'event' in English implies a beginning and an end.) Swain means, presumably, events that are not placed (or not consistently placed) in relation to others in a sequence. Furthermore, while Rudder (1993) agrees with Swain in saying that Yolngu ancestral events are timeless, his analysis of Yolngu temporal constructs shows how they are located in what he calls Yolngu 'temporal space'. In Rudder's analysis, time or rather 'temporal space' just is the conceptualisation of past, present and future. It does not imply linear time in the sense of a numerical time-line. Swain can, of course, save his analysis by invoking the Lévi-Strauss stratagem, and assert that linear time has arisen in the context of social change, safe from the possibility of testing his hypothesis against any evidence (Swain 1993:23).

Until the adoption of clocks and watches Yolngu did not have a concept of time in the abstract like Newton's 'absolute time', which can be measured, that flows, that can be 'saved' and 'wasted'. Rather, they locate events in 'temporal space' with reference to repeated events of the natural world, and the succession of unique (human) events. The repeated events, what we might call 'clocks' and 'calendars', include the position of the sun in the sky, often relative to a person's body (for example, *l̲a:y-bilyunaray*, 'sun turning to strike the temple'), elaborated by specifying qualitative attributes of a position such as the direction of shade. The sun's positions divide the day up into a series of periods each with a particular set of qualities. Adverbs of time denote relative distance from the present in temporal space (Rudder 1993:244-5).

Yolngu recognise six or seven seasons of the year, each with a name and characterised by the presence, absence or quality of rain and storms, the prevailing wind direction, plant growth, temperature, and food available. The year does not have a specific beginning and end as in Western concepts of time – when recounting the cycle of seasons a person will begin with the current season (Rudder 1993:252).[4]

Yolngu astronomy is not very elaborate, but certain stellar events do provide reference points, such as the 'turning' of the Milky Way (*baḏurru*), the position of the Pleiades and Orion, and the rising of a series of 'morning stars' (*baṉumbirr*). The phases of the moon provide another cycle or series to which Yolngu relate past or anticipated events, with reference to its position in the sky, phases, and I might add, position in the sky at sunset (Rudder 1993:255).

What seems of importance here is that events are placed in a space of relative distance from the present, not at a position in an absolute objective series. Ancestral events occurred 'long ago', in 'far off' times, but this is to place events relatively distant from the present, neither within a definite period or in a timeless zone of 'everywhen'.

However, actions or events can also be placed in temporal space relative to unique series of events or key kinds of person. These are not like the universal numerical clock and calendar, but vary from reflections of collective experience to more idiosyncratic experiences. The events and persons of collective experience include:[5]

revivalmirriy	at the time of the Christian Revival
mitjinmirriy	at the time of the Mission
womirriy	at the time of the World War II
manggathamirriy	at the time when the Macassans used to visit Arnhem Land

I would add:

wangarryu	at the time of the *wangarr* ancestors[6]

These temporal referents vary in their precision and duration, moreover they overlap: *womirriy* occurred during the Mission era.

To these we could add the presence of a well-known person such as a mission superintendent or township manager – 'in *ba:pa* Sheppey's time'. At the more idiosyncratic end are events in a person's or a relative's life history which serve as temporal referents, such as the person's circumcision (male), when the person's first child was born, when the person's grandmother's first child was born, and so on.

Intermediate along the scale of individual to collective experience are the first people with a unique patrifilial group identity, and who in some myths are coevals of *wangarr.* Their personal names are 'elbow names' (*likan ya:ku*) which have been adopted recently as European-style surnames. Rudder places a position in temporal space such as 'when the founder of the group was alive' at the 'mythological' end of the temporal continuum (1993:271).

It is not so much, then, that there was a mythological, ancestral *period* to which 'myths' refer, but rather that there existed particular kinds of being, *wangarr* ancestors, whose presence and actions, very distant from the present, are recounted in stories, including stories of the kind that Morphy (1990) calls 'myths of inheritance'. Certain actions and events are placed in relation to the presence of *wangarr* by means of the agentive suffix – *wangarryu*, 'ancestral times'.

The tense of verbs used to denote past events is relevant here. According to Lowe's analysis of Gupapuyngu, the 'today past' which used the tertiary form of the verb (for example, present tense *bathan*, 'cook', tertiary form *bathara*) is used also to refer to the distant ancestral past. This contrasts with the 'yesterday past' used to refer to recent events. Rudder recasts this distinction: the tertiary form of the verb refers to events that are not precisely located in temporal space – the indefinite past, contrasting with the definite past which locates events more precisely. The indefinite past is used when recounting ancestral events in 'myths'. Rudder relates this contrast between definite and indefinite past tenses to the degree of precision in the temporal location of events. The closer events are to the present, the more precisely Yolngu discourse locates them; the further away, the less precisely.

Ancestral events, 'long ago', are recorded in the indefinite past. Even so, Rudder reports recording a story about events which seem to have occurred in 1942 (presumably events of World War II) 'couched in mythological terms and given an indefinite temporal location'. It was a myth of inheritance rather than a myth of creation in terms of the contrast drawn by Morphy (Morphy 1990; Rudder

1993:274-5). There are two possible interpretations of this; either 'mythological' events need not be very distant – 1942 was the time of the current senior generation's childhood or adolescence, or *womirriy* (the time of World War II) has become 'long, long ago'. Either way, Yolngu temporal space is telescoped by the standards of Western absolute time.

I agree with Swain (1993) and Morphy (1990:314) that Yolngu time is not 'cyclical', but it is clear that Yolngu place events in temporal space relative to unique successions of events, measured against the course of a person's life, the succession of generations, major events, or the presence of persons of which some are placed in a remembered relation of succession. Adverbs of time measure the relative distance of presences and events from the present. The quality of ancestral times derives from the kinds of beings present in Yolngu places, and their extraordinary powers. Ancestral stories do not recount events in a timeless domain of 'Dreamtime' or 'Dreaming', but locate the events at a position in temporal space relatively distant from the present. By 'Western' standards, ancestral times are relatively close. This location allows for interaction between *wangarr* and people placed at a less distant position, for example in genealogical time/space. Human ancestors, capable of being placed at the apex (or root) of genealogies, interacted with *wangarr*, and the recently dead have returned to the domain of *wangarr* in the waters. Manifestations of *wangarr*, including spirits of the dead, may still be encountered by the living. Exegeses of the significance of places interweave the actions and traces of *wangarr* ancestors and close relatives in the same register; ancestral stories can be deployed as analogies of the actions of known persons, living or dead; and stories about recent events may be cast in an idiom similar to that of stories about *wangarr* ancestors. A categorical distinction between 'myths' and other stories does not, therefore, seem warranted. When describing and classifying Yolngu narratives it would be preferable to follow Yolngu practice and add the appropriate adjective to the word 'story'.

The continued presence and powers of *wangarr* under the waters and in the land complicates the picture. It is tempting to go along with the 'everywhen' argument and suggest than since *wangarr* remain, the temporal space of *wangarryu* still prevails in parallel with the domain and temporal space of human affairs. The use of the temporal markers 'far off' and 'long ago' lead me to resist that temptation. More consistent with Yolngu ontology is that as personages and events become more distant (and here the perspectives of the youngest generation differs from that of the oldest) so that their potential for becoming 'ancestors' increases, marked when they become 'inside' and then 'outside' subject matter of songs and ceremonies. Of course, insideness – deep in waters, or set apart by secrecy – implies an analogous relationship in spatial and temporal distance.

Conclusions

The discussion has moved rather far from the old Airforce road. But its direction suggests that the road may not, after all, have been an inappropriate candidate as a legitimate boundary marker in RM's eyes. The traces of human action are inscribed in a similar frame to ancestral action; and the events of World War II may now be distant enough in Yolngu temporal space for the possibility of various kinds of interaction with *wangarr* times; unless the younger generations are adopting Western absolute time, in which case *wangarr* will become truly mythological.

These remarks about Yolngu concepts of time and types of narrative have implications for another issue involving Aboriginal communities and mining in the region, and that is the Coronation Hill case (Merlan 1991, this vol.; Brunton 1992; Keen 1992, 1993). Elsewhere (Keen 1993) I suggested that certain arguments put forward in support of the goldmine at Coronation Hill, and against the registration of the area as a sacred site under the *Aboriginal Areas Protection Act* of the Northern Territory, at once recognised the potential binding force of certain Aboriginal beliefs and norms on non-Aboriginal as well as Aboriginal people, while at the same time restricting the scope of that force.[7] One of the key issues was the interpretation by Jawoyn people of recently discovered gold ore as ancestral essence or substance. Francesca Merlan (1991) pointed out some of Ron Brunton's rhetorical moves in his critique of the anthropological research for the Resource Assessment Commission during the Coronation Hill Inquiry, and of the case against the mine going ahead (Brunton 1991a, 1991b, 1991c). To be genuine, (Aboriginal) traditions must be unrelated to interaction with outside issues and agents; because some of its effects are negative it should not coexist with modernism; and tradition is acceptable as such only if it can be shown to be completely unchanging over a long period of contact with a settler society (Merlan 1991:349-51).

Since the interpretation of newly discovered features has been reported as an aspect of Aboriginal religions, I asked, why should Aboriginal interpretations of newly discovered minerals as ancestral substance be discounted, given that sacred sites and other legislation in principle accords Aboriginal religious beliefs recognition (Keen 1993:353)? In effect, I suggested, beliefs and norms were placed in a containing 'frame':

> Assertions and modes of action are categorized as 'religious beliefs,' 'traditions,' or 'heritage' characteristic of an ethnic category with special status, and thus worthy of respect ... Assertions that appear to be contingent expressions of individual opinion rather than the 'authentic' attributes of an ethnic category are deemed irrelevant. An assertion must be recognizable as a belief characteristic of that category, rather in the way that 'Aboriginal art' must be recognizable not as the work of a particular individual but as characteristically 'Aboriginal.' 'Beliefs' are reified into fixed cultural traits, and responses to changing conditions tend to be ruled out. [Keen 1993:353]

If the debate over Aboriginal beliefs in the Coronation Hill case had been conducted outside such a frame, presupposing that the truth-value of Jawoyn assertions about ancestral substance and power was (directly) relevant, then the terms of the public debate would have been very different. Opposing parties would have attempted on the one hand to establish that the gold ore really was the essence of the ancestor, Bula, or the substance of his wives, or on the other to show that it was not (p.353). (As a matter of fact, some critics of Jawoyn opposition to the Coronation Hill mine in the tabloid press did argue in just these terms.) There is also a spatial aspect of containment within a frame of recognition that insists on traditionality, in the form of the discrete, clearly demarcated and bounded 'sacred sites'. (The very size of the sacred site including Coronation Hill, proclaimed by the Aboriginal Sacred Sites Authority (as it then was), caused dismay.)

Let me now bring the earlier discussion of the relevance of the categories 'myth' and 'history' to bear on these aspects of the debate over Coronation Hill. The containment of Aboriginal religious discourse and practice within a frame of traditionality, together with the spatial containment of associated places, is consonant with an insistence that the only legitimate binding beliefs are just those of the world of 'myth' rather than 'history', having relevance to distinct kinds of 'heritage' places in terms of a fixed traditional world.

In contrast, the discussion of Yolngu categories of story and concepts of time in relation to the old Airforce road suggests a world in which ancestral action, powers and substance are not so temporally or spatially contained. Yolngu do not draw hard and fast boundaries between the world, time and narratives of the *wangarr* ancestors and those of the recent past and present, or of the 'everyday'. Nor do they draw boundaries between 'sacred sites' associated with unchanging traditions about *wangarr* ancestors or other spirit beings, and everyday, profane places with no such associations.[8] Rather, in a rapidly changing world of interaction with non-Aboriginal people, Yolngu 'inscriptive practice' continually interprets action, events and places in terms of very proximate ancestral action. For them, ancestors are not creatures of 'myth'.

Notes

1. 'Boundary' seems to have two senses here; in its first usage it seems to denote 'place' or perhaps 'area', but in the second usage I understood it to have more the Australian Standard English meaning.
2. This passage is not verbatim, but paraphrased from field notes.
3. Sutton links the history-making of urban Aboriginal people to myths of the Dreaming. The Aboriginal person of these histories is 'not merely the survivor but the embodiment of the scarifying process of conquest, dispossession, resettlement, missionisation and welfareism'. The contemporary white Australian can be defined and identified as the historical coloniser. This history-making resembles 'traditional' Aboriginal approaches to myth and remembered events in not being committed to the canons of empiricism, and in having similar sociological usages (1988:261).
4. The names for the seasons are: *gu<u>n</u>mul* (roughly January to February), *myaltha* or *ba:rramirriy* (March to April), *mi<u>d</u>awarr* (May), *dharathara* (June to July), *rrarrandharr* (August to September), *wulmamirr* (October to November), and *dhulu<u>d</u>ur* (November to December).
5. In Yolngu, temporal adverbs of this kind are formed by adding the ergative/instrumental/causative suffix *dhu/thu/yu/-y* to a noun denoting the person or event that is taken as the reference point (see F. Murphy 1983:39). Reference to a season, for example, can be by its name, such as *mi<u>d</u>awarr*, or a phrase including a nominalised verb (the tertiary form)

plus the agentive/instrumental suffix, as in *ngatha galadare'yunaray*, 'yam leaves change colour' (Rudder 1993:267). Demonstratives also have the agentive form, as in *dhiyangu bala waluy*, 'today'. Adverbs of time can also be formed by a noun plus the Proprietive suffix *mirri* , 'having', plus the agentive suffix (this is the Gupapuyngu form; it varies among dialects). Thus the season of the strong west wind in March to April is *ba:rramirriy*, 'when the strong west wind blows' or *miyaltha* (Rudder 1993:255).

6. While I have recorded this usage, Rudder says that he has not, but that it is not inconsistent with the other temporal referents.

7. These arguments included the claim that an Aboriginal group has a traditional right to the country as well as a long association with it; that beliefs should be current and associated with ongoing ritual tradition; the beliefs should be held by a majority of the group, and be held sincerely and consistently; and they should be unchanging (Keen 1993:353).

8. Yolngu do distinguish 'outside' or 'ordinary' places from 'inside' *madayin* places, but this is not to say that they have no ancestral associations.

9

CHANGING VIEWS OF PLACE AND TIME ALONG THE OK TEDI

Stuart Kirsch

> The dialectical oppositions between place and space, between long- and short-term time horizons, exist within a deeper framework of shifts in time-space dimensionality that are the product of underlying capitalist imperatives to accelerate turnover times and to annihilate space by time. The study of how we cope with time-space compression illustrates how shifts in the experience of space and time generate new struggles in such fields as aesthetics and cultural representation, how very basic processes of social reproduction, as well as of production, are deeply implicated in shifting space and time horizons. The production of spatio-temporalities is both a constitutive and fundamental moment to the social process in general as well as fundamental to the establishment of values. And that principle holds cross-culturally as well as in radically different modes of production and significantly different social formations. [Harvey 1996:247]

This is an ethnography of loss and of innovation. My subject is the response of the Yonggom of Papua New Guinea to the impact of the Ok Tedi copper and gold mine on their environment. (For locations see Map 9.) In production since the mid-1980s, the mine releases 30 million tons of tailings and 40 million tons of waste material into the local river system annually (Ridd 1997). Pollution from the mine has resulted in deforestation, the destruction of garden land and sago swamps, the disappearance of birds and other wildlife, and unknown chemical hazards. In this chapter, I examine the consequences of these changes for the Yonggom and their relationship to the surrounding landscape.

This discussion is intended to contribute to debates about the impact of large-scale resource development projects on indigenous communities, which can destabilise the conditions through which

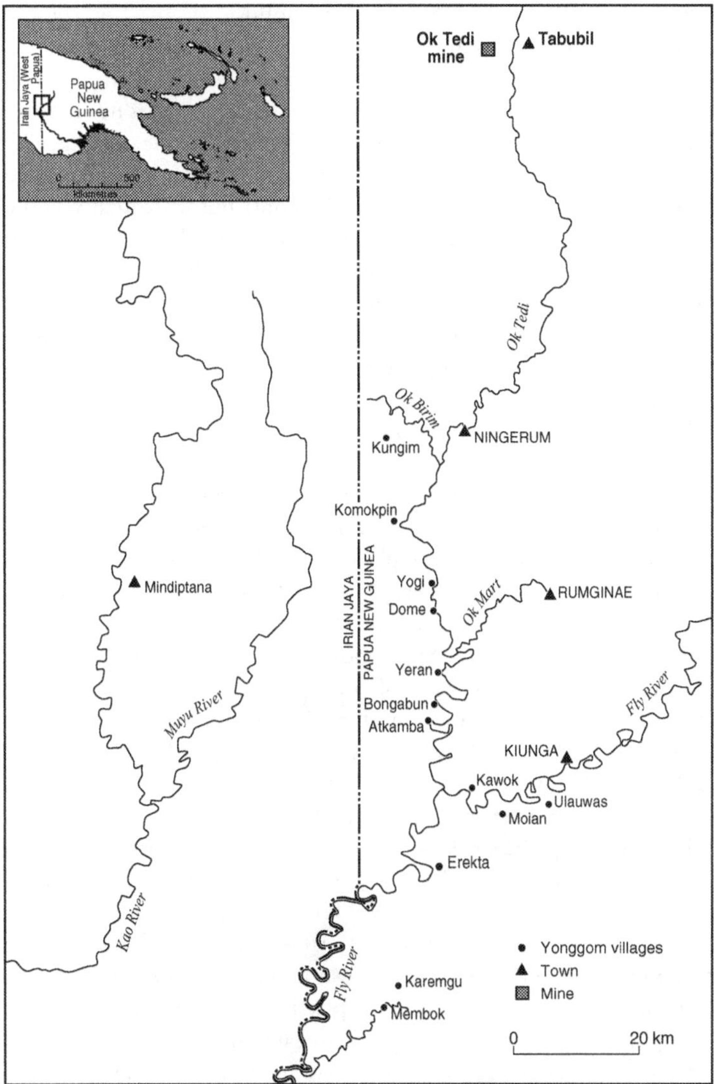

Map 9: The Ok Tedi mine and Yonggom villages in Papua New Guinea

these societies reproduce themselves.[1] Euro-American assessments of environmental impact, which emphasise economic costs and benefits, biological continuity and aesthetic impairment, are insufficient for analysis of Yonggom responses to pollution caused by the Ok Tedi mine.[2] It is necessary to account for their experience and understanding of local landscapes as well as the kinds of social processes they build from them (see Strathern 1988; 1997:7). In my inquiry into these practices, I take as my starting point the idea of 'place' as articulated by Steven Feld and Keith Basso (1996) in their edited collection *Senses of Place*. Whereas the contributors to that volume emphasise the ethnographic task of 'set[ting] forth as accurately as possible what being-in-place means' (Casey 1996:15), my focus is on how such basic understandings may change.[3] I describe how the deterioration of local landscapes has led to an ontological shift in Yonggom perspectives on place and time, and trace the conceptual and epistemological dimensions of these changes.

My argument thus attempts to bring together political economy and phenomenology. In *Justice, Nature and the Geography of Difference*, David Harvey (1996:315) argues that there is a dialectical relationship between capitalist or modernist space-time and the phenomenological experience of place. Nonetheless, he maintains that the interconnections between places established through the global economy requires that 'we should start to think of these arguments not as mutually exclusive but oppositions which contain the other' (ibid.:315). Alternative ways of conceptualising environmental relationships can enrich contemporary debates about the costs of economic development. This analysis of Yonggom ideas about place and loss is therefore intended to amplify the power of indigenous ideas as counter-narratives to the prevailing discourses of capital.

I also take up the problem raised in the debate between Alfred Gell (1992a) and James Weiner (1995b) regarding the differences between art and technology. Weiner argues that Melanesians see art and magic as means of revealing those aspects of the world that make creation and production possible. He reiterates Heidegger's call for a non-productionist metaphysics as an alternative to the metaphors of production ingrained in capitalism. This argument is central to my discussion and analysis of how the Yonggom perceive their landscape as possessing hidden possibilities that can be revealed through enchanted means. Their relationship to the unseen dimensions of their world has been compromised by the mine, leading them to conceptualise new modes of 'being-in-place.'

I begin by describing how Yonggom actions within the landscape are incorporated into narratives that link biography and history to place. Pollution from the mine has transformed local places along the river, disrupting the spatial form of these narratives. In response, the Yonggom have begun to emphasise chronological frames of reference. These changes affect their relations with the other beings with whom they share the landscape, which has implications for their religious beliefs and practices. Environmental degradation challenges their perceptions of environmental security and risk as well. By forging alliances with conservationists and other international activists, however, the Yonggom and their neighbours have been successful in challenging the mine through protest and legal action. These events influence Yonggom debates about the future and the nature of change in the new millennium. In conclusion, I discuss the relationship between the enchanted technology of the Yonggom and capitalist understandings of productivity, and argue that political-economic analyses should take account of alternative modes of 'being-in-place.'

Place

When I began research among the Yonggom in 1986, local histories were mapped onto the landscape, with places metonymically representing experiences in a person's life. In the course of a lifetime, the Yonggom acquire detailed knowledge of their land. They learn the location of useful trees and plants. They know the fruit stands where birds come to feed and how to find pig and cassowary tracks. They know where to catch fish, crayfish and turtles. The Yonggom also transform the landscape through its use. They maintain a network of trails, camping places and catchments for drinking water. They control sago stands through selective harvesting and replanting. They clear and burn the forest for swidden gardens. They plant fruit trees in these clearings as well, which reach maturity long after their gardens have been abandoned. They fell trees to make canoes and to build houses. Gradually the landscape is transformed so that it reflects the biography of its owner.

Lineage names signify physical places as well as social groups. When a lineage divides in two, the new groups are identified by residence, so that Miripki lineage, for example, may become Miripki-kubunun, Miripki-on-the-waterfall, and Miripki-yumkap, Miripki-by-the-bananas. Places may also figure in the formation of *won* reciprocal nicknames that formalise dyadic relationships (Kirsch 1991; cf. Schieffelin 1976:56-57). Land and identity are coincident.

The Yonggom also read the traces of others on the landscape. When walking through the forest, people identify the hunting party that camped at the junction of two paths or the man who planted a tree that matured after his death. They refer to the owners of a sago stand, the hunter who shot a flying fox in the arms of a breadfruit tree, or the location of a male initiation ceremony (*yawat*). Places are linked with human activity; as Edward Casey (1996:27) suggests: places 'not only are, they happen.'

These inscribed histories extend beyond the lifetime of the individuals whose activities produce them. What Weiner (1991) has written about the Foi holds true for the Yonggom as well: 'the life of the deceased is depicted as a series of places that belonged to him, or that he inhabited.' After a death, people mourning the loss of a relative or a close friend may refuse to leave the village for long periods of time to avoid confronting the memories of the deceased that continue to echo through the forest.

Throughout southern Papua, biography is conceived of as itinerary, as Feld (1996:113) has observed for the Kaluli. The conjunction of place and event provides the primary mode of remembering and representing past experience. Yonggom narratives about the past are organised in spatial terms: where one was born, lived as a child, went hunting or made sago, and so on. Individual lives are seen as a series of movements across the landscape. Memories of the past are linked together like campsites along a trail, organised by physical proximity rather than chronological sequence. Bumok Dumarop told me her life story this way, describing how she moved from the Muyu River to a place called Dunumbit, then to a settlement at Mokbiran and from there to the area along the border, where she spent several months helping an uncle prepare sago for a feast. Later she moved across the border, settling in a village beside the Ok Tedi River in Papua New Guinea.

While time flows through Dumarop's narrative, she tells a story of movement between places. These are not arbitrary landmarks, but the locations of meaningful activity. Yonggom narratives make the nexus of person, place and event come alive in public consciousness and memory.

By virtue of being related as journeys, these Papuan spatial ontologies always incorporate the passage of time, as Andrew Lattas (1996) emphasises in another context. Alfred Gell (1992b) compares two modes of representing time, the first of which organises events into a linear sequence of past, present and future. Alternatively, events may be categorised in relation to each other; that is, whether one occurs prior to, simultaneous with, or subsequent to the other. Whereas chronological time emphasises linear relationships, Yonggom narratives about place stress relative position rather than absolute order. Place and time are incorporated into both modes of representing experience, but with differential emphasis.

Howard Morphy (1995:188) makes this point in his analysis of Yolngu ideas about place and time:

> Place has precedence over time in Yolngu ontogeny. Time was created through the transformation of ancestral beings into place, the place being for ever mnemonic of the event. They 'sat down' and, however briefly they stayed, they became part of the place for ever. In Yolngu terms, they *turned into* the place. Whatever events happened at the place, whatever sequences they occurred in, whatever intervals existed between them, all becomes subordinate to their representation in space ... What remains is the distance between places rather than the temporal distance between events.[4]

For the Yonggom, temporality is also subordinated to the landscape, which is the form taken by history. Travel through the rainforest reveals places that were sites of meaningful events in the past. The future is also conceptualised in terms of the revelation of possibilities otherwise concealed within the landscape, as I describe below. The resulting notion of place has an unseen dimension that is central to Yonggom ideas about productivity and the manifestation of opportunity.

Loss

Today, when walking near the Ok Tedi River with a friend, it is difficult to locate the places where we once shared a meal or went swimming. Where towering trees once stood there are only ghostly tree trunks and the creeks have all been buried by sand. This transformation of the landscape not only produces spatial disorientation, it also displaces memories of the past. A young woman who remembered making sago with my wife expressed dismay that the sago swamp where they once worked together was now dried up and filled with sand. Memories previously anchored by the landscape have lost their mooring.

During a recent visit to the Ok Tedi, I walked with Buka Nandun to where his mother used to make her gardens. Until recently this was an island in the Ok Tedi River known as Dutbit, the site of fertile land that bore fruit without fallow. Tailings from the mine have covered the island, which has been joined to the mainland. No one gardens here any longer. The few mature trees that remain lean precipitously. Leading me to an area through which he has walked many times before, Buka lost his way in a thicket. Craning his head right and left, he searched in vain for a landmark that might show him the way out. Pollution has erased all traces of the past.

What is the meaning of these empty places? What kind of places are these? Casey (1996) describes memory as fixed to place and places as the repositories of memory. With the destruction of local landscapes, memories are severed from the site of their creation: Dutbit Island is not just an empty place, but a scene of loss.

In life history interviews, women in the villages described how their lives have changed since production began at the mine. They expressed feelings of *mimyop*, of sorrow and loss, at the way that the landscape along the river has been disfigured.[5] Duri Kemyat from Yogi village, a woman in her mid-fifties, described this transformation in the stylised form of a lament, the speech genre associated with bereavement:

Before the river was not like this;
it makes me feel like crying.
These days, this place is ruined,
so I feel like crying.

Where I used to make gardens,
the mud banks have built up.
Where I used to catch prawns and fish,
there is an empty pool ...
So I feel like crying.

Before it wasn't like this.
We had no difficulty finding garden food and wild game.
We had everything we needed.
Now we are suffering and I wonder why.

The narrative coupling of place and event is frequently associated with the expression of loss in Papuan societies. This may take the form of the loss of a human life or the disappearance of the traces of that person's life in the form of abandoned gardens or house sites that have been reclaimed by the forest in the absence of the human intervention required to keep these places open and productive. These losses are revealed through movement across the landscape, whether a physical journey, the biographical accounting of one's movement between places, or a metaphorical journey, as in the Kaluli *gisalo* (Schieffelin 1976, Feld 1982, Weiner 1991, Munn 1996). The actions of the other beings with whom the Yonggom share the landscape may also call attention to loss. For example, the song of the bird *on kuni* (hooded butcherbird) evokes memories of deceased relatives. Flowering sago palms are associated with feelings of sorrow and loss as well. These trees only flower once after twelve to fifteen years of growth. They must be harvested before flowering as the efflorescence consumes the bulk of the tree's edible starch. A flowering sago palm, because its starch has gone to waste, evokes memories of relatives too old or too frail for the labour required to process sago, or who have already died.

In these cosmologies of emplacement, aspects of the landscape are revealed by human and non-human movements and events, such as a journey between places or bird song, evoking a sense of loss that is generically associated with place. The resulting association between memory and place constitutes their experience of loss and has shaped the Yonggom response to the mine's destructive impact on their landscape. Kemyat's moving elegy to place and loss – 'Where I used to make gardens/the mud banks have built up/Where I used to catch prawns and fish/there is an empty pool … /So I feel like crying' – reflects their understanding of the landscape as the embodiment of history and therefore the medium through which the experience of loss is made explicit.

If the 'power of place is that it counters the process of entropy' (Casey 1996:25), then the challenge to the Yonggom is how to retain their memories in the wake of destruction. Peter Munz (cited in Hughes 1995:3-4) compares the separation of the structures of time from human experience to filleting a fish, which leaves behind 'something like a mollusc – a wobbly, still undifferentiated mass … deprived of its time skeleton'. Yonggom narratives have their own sense of time, which is mapped onto concrete places and represented as movement across the landscape. These narratives are the 'time structures' that give form to their experience of time and place. In response to the destruction wrought by the mine, Yonggom narratives have been reworked to incorporate alternative modes of temporality.

Time

Yonggom narratives of past experience are increasingly structured in chronological terms. Some of these chronologies were whispered furtively to me, much the way that people once divulged the *waruk* magic names of objects. *Waruk* represent the true names of things: of animals, of lightning, of the ground itself. Knowledge of *waruk* gives its owner influence over the things to which they refer. People only revealed *waruk* names to me in confidence; when alone they sometimes introduced important myths by reciting a string of *waruk* names that authenticated their story, much like totemic names are recited in the Sepik (Harrison 1990).[6] These chronologies delineate the gradual incorporation of the Yonggom into broader social, political and economic systems. They are structured by key historical events, such as the arrival of the first missionaries, the founding of the town of Kiunga, the establishment of the village and the beginning of production at the mine. These important dates are treated analogously to *waruk* names that bring about change by virtue of being spoken aloud; the implication is that knowledge of key historical dates confers power over the processes of change. Kewo Yang, a retired policeman, insisted that I record the chronology of the mine, from the early days of prospecting in the 1970s to the problems after production began in 1984.

People have begun to recount their life histories in chronological form as well. Some of these stories begin by referring to the places where the events occurred, but shift to a temporal framework after 1984, the year that a cyanide tap at the mine was left open overnight, killing thousands of fish, crocodiles and turtles along the Ok Tedi. Andok Yang, a woman from Dome village, described how circumstances changed after the cyanide spill: 'People wondered what would happen next. This was the beginning of the sand banks that later covered our gardens along the river. By 1986, the plants and trees along the river began to die. Their leaves turned yellow and fell off. Gradually the effects of the mine spread into the swamps where our sago palms grow and into the surrounding forest. The creeks filled with mud, killing the sago trees. The sand banks along the river grew higher …'

The shift from a predominantly spatial representation of past experience to temporally organised narratives is one of the conceptual consequences of environmental degradation caused by the mine.

It implies a fundamental change in how the Yonggom view their relationship to the world around them. With reference to past events becoming increasingly structured by abstract chronologies, their remembrance of things past is no longer linked to their surroundings. New biographies are being imagined that have the potential to separate experience from place.

In the anthropological literature on the politics of time, calendars and clocks are usually described as instruments of domination through which political elites establish control over the broader body politic (Bourdieu 1977, Rutz 1992). Henry Rutz endorses this view: 'A *politics* of time is concerned with the *appropriation* of the time of others, the *institutionalization* of a dominant time, and the *legitimization* of power by means of the control of time' (ibid.:7). Katherine Verdery (1992:37) emphasises the political context of encounters between the 'bearers of nonwestern or non-capitalist temporalities ... and the new organizations of time brought to them by capitalist commodity production.' She defines the politics of time as a political struggle between 'social actors who seek to create or impose new temporal disciplines ... and the persons subjected to these transformative projects' (ibid.:37). Most importantly, she observes that 'struggles over time *are* what construct it culturally, producing and altering its meanings as groups contend over them' (ibid.:38). Social consciousness of time is the product of power relations.

The Yonggom have turned to new forms of chronological and calendrical reckoning in response to environmental degradation and its threat to local models of place and time. Although facilitated by education and literacy (Kulick 1992, Smith 1994), these changes suggest other means by which capitalist modes of production can influence local ideas, both by disrupting local constructions of place and by providing alternative ways of conceptualising experience. Temporal innovation, including a linear model of events, provides the Yonggom with new ways to make reference to the past.

These new forms of temporality have two significant implications for how the Yonggom conceptualise their relationship to the landscape they inhabit. Whereas their experiences once overlapped completely with the lives of the animals and other beings with whom they interact, the decoupling of time from place emphasises ontological differences between humans and the other inhabitants of their shared landscape. While the Yonggom otherwise view the future as an aspect of place that can be revealed through enchanted means, these new models of time separate the present from the future.

The nature of communication

Central to Yonggom notions of place is the presence of animals and other beings that are conceived of as having powers of agency comparable to that of people, much as Don Gardner (1987:170) has described for the Mianmin:

> The central feature of these animistic beliefs is that the natural and social worlds, the worlds of things and persons, are subsumed under a single and familiar scheme of explanation.
>
> The crucial aspect of such schemes is the all-persuasiveness of agency as a principle of the functioning of the world and explanation of events of all kinds (ibid.:162).

According to this perspective, agency is not restricted to humans, but extends to the other beings with whom the Yonggom share the landscape, a quality that Tim Ingold (1996:129) describes as 'interagentivity.'

Yonggom knowledge about the world around them is structured by communication with the other inhabitants of their landscape. They are adept at recognising many bird species by their calls, such as the ear-piercing cry of *on kawa*, the sulphur-crested cockatoo, or the nasal honking of *ono*, the greater bird of paradise. Birds indicate the time of day by their calls and movements, the seasons by their appetite for ripening fruits and the weather by migrations during periods of drought. Birds mark sacred time as well; during male cult ceremonies, everything grows quiet as the birds are said to cease singing altogether. Birds identify themselves or speak in Yonggom language, like the large-tailed nightjar known as *on dokdok*, which calls out *'dokdok dokdok, dokdok',* or the bird the Yonggom call *on kam*, which says *'kwi, kwi, kwi,'* meaning 'like that, like that, do it like that.' Birds also signal impending misfortune by appearing at the wrong place and time: an owl calling in the village at night means that a sorcerer is on the prowl.

Other animals communicate with the Yonggom as well: cicadas crying out before dusk tell people still in the bush to hurry home. Natural phenomena also convey messages: the golden light of the sunset known as *dep aron*, named for the colour of the marsupial *bandep*, warns of death. The appearance of animals in dreams provides insight into the future. A dream about a crowned guria pigeon indicates that a *kumka* assault sorcerer will come to the village. A kingfisher's appearance in a dream is a harbinger of illness. Pigs and cassowaries have inverse relationships with men and women in the dream world: a pig's appearance in a dream means that a man will visit the village, whereas to dream of a woman signals an opportunity to hunt for cassowaries. The referents of these signs are not limited to natural events, such as the danger posed by rising rivers or a propitious time to plant. They also provide critical social information, including warnings, predictions and the indication of opportunities upon which people can act (Wagner 1972:55-84). As Gardner (1987) suggests, the laws governing the physical world are seen to guide human affairs as well.

Communication between the Yonggom and the other inhabitants of the landscape moves in both directions. The Yonggom use magic spells to convey their intentions and desires to the birds, fish and other animals with whom they share their landscape. The following invocation is used when hunting guria pigeons:

on kurim	guria pigeon
kup ku kirot mene	you come here quickly
menip kop, weetmore wana	come forward so that I can see you
ne ku munggi bopman	I am starving to death
kowe, kwi	so, do it like that

A similar spell is used to catch fish:

on yip, ku ne doberan ki	you fish, I am waiting
kirot, yaro minime!	quickly, you must come!
menip kop	all of you come
monbe, monbore	shoot, I am shooting
de ambioom wana	and then I am going home

These hunting spells urge animals in the rainforest and the river to come forward so that they 'may be seen.' This is sometimes expressed through the use of the adjective, *ayimamip*, which combines the verb stem *aye*, meaning to hit, strike or shoot, and the suffix -*mamip*, which like the English -able and -ible implies capacity, fitness or worthiness. Thus *ayimamip* means 'possible to shoot' or 'within range.' The contrasting term is *akmimamokban*, which means 'hidden' or 'unseen,' and is composed of the verb stem *akme*, to see, the negative marker *ban* and the infix *-mamok*. *Mamok* is the negative form of *mamip*, implying the lack of capacity. For example *animamokban*, meaning 'inedible,' is a parallel construction using the verb stem *ane*, to eat.

These spells thus preside over the transition of things that are hidden or unseen, known as *akmimamokban*, to that which is visible or *akmimamip*.[7] In this latter state, that which was concealed is now manifest, creating an opportunity upon which one can act, much like a message from the natural world. This magical strategy operates as a discovery procedure, a technique of elicitation, a form of revelation.[8] The intended event, in this case a successful hunt, involves a process of manifestation in which the seen and unseen aspects of the world are brought together. Production in the Yonggom view entails the revelation of the future through the conjunction of these two aspects of reality.

These spells also exemplify Weiner's call for a non-productionist metaphysics. 'What if the world of production and making, of consumption and controlling was only elicited,' Weiner asks rhetorically, 'what if it were the reflexive by-product of something else, like magic and art?' (1995b:33). Yonggom hunting magic illustrates the assertion that Melanesians view magic and art as foundational acts. Productivity is contingent upon magical techniques of elicitation that 'bring-forth' that which is otherwise hidden or concealed.[9] These magic spells reveal the hidden potential of place.

Yonggom hunting spells do their work, however, by the magic of communication: they compel, they cajole and they persuade.[10] They give instructions, such as *mene*, 'you come,' which may be emphasised by the use of the imperative *minime*! or 'you must come!' They persuade by means of exaggerated claims, such as *ne munggi bopman*, which literally means, 'I am starving to death.' They use temporal adverbs like *kirot*, or 'quickly,' to establish a sense of urgency. This may be underscored by use of the contrasting verb, *dobere*, to stand and wait. *Dobere* implies impatience as well, as in the formulaic utterance: *ku ne doberan ki, kirot, yaro minime*! which means: 'I am standing here waiting, so you must hurry up and come!' The spells also refer to the completion of the intended act by a shift in tense from the present to the transitive (for example, 'shoot, I am shooting') and by reference to events that come afterwards, *de ambioom wana*, which means, 'and then I am going home.'

These spells are the vehicle through which the hunter attempts to impose his will on the other beings with whom he shares the landscape. They depend on the assumption that people and animals form a single speech community. Relying as they do on persuasion, the spells acknowledge the agency of these other beings as well.[11]

Through their messages and their appearance in dreams, the other beings with whom the Yonggom share the landscape invoke a notion of time in which the future is made manifest in the present. Much like events in the past are revealed through movement across the landscape, the future is an unseen aspect of the present that may be revealed through enchanted means. Yonggom notions of time are structured by the same opposition between the seen and the unseen that organises their sense of place. The Yonggom do not have a linear or evolutionary sense of time, but a view in which the future appears in the present much like the hidden is made manifest in a clearing through magic. The future is a hidden aspect of reality that can fill up the present (the now) with signs of itself, much like magic fills up the present (the here) with things. The future is therefore present but not visible until it is made manifest through the enchanted agency of the Yonggom or the other beings with whom they share the landscape.

Yonggom religious beliefs are animated by the presence of the beings with whom they share the landscape: what are the consequences for their religious imagination when there are no longer any fish in the river, no birds overhead and no game nearby, as is the case along the Ok Tedi River today? Whereas Yonggom hunting magic works by revealing what is concealed by the rainforest and the river, pollution from the mine has destroyed the hidden potential of the landscape. Magic has lost its audience. The forests have grown quiet and the dialogue the Yonggom once had with the animals around them has all but ceased.

Yonggom ritual and myth are populated by birds, fish and other animals. Yet symbols are only powerful repositories of meaning when the referents are familiar. Without exposure to these animals in their natural habitats, Yonggom rituals may lose their capacity to communicate insights about the human condition or to solve the dilemmas that people face. Under such circumstances, myth may degenerate to the level of amusing folk tales, known as *stori tasol* in the Pidgin. Even if the message of the myth remains clear, its characters no longer share the landscape with the Yonggom. Similarly, some people now see ritual as something that tricks rather than enlightens them.[12] Although hunting magic may still be effective away from the river, where the land is not affected by the mine, its local collapse foreshadows a time in which magic will no longer have any efficacy. These narrative conventions are only powerful and compelling because they have effects in the material world, and when that world alters profoundly, the power of speech itself may become attenuated.

From security to risk

Yonggom subsistence once depended upon their knowledge of a wide variety of plant and animal species. They traded with their neighbours, but were largely self-sufficient in terms of food production. As they became more active participants in the world economic system, they began to exchange their labour, as well as a share of their garden, forest and riparian products for commodities produced elsewhere. Until recently, however, their environment provided them with both subsistence and security.

This aspect of the Yonggom relationship to the physical environment has changed in two fundamental ways. First, they are haunted by a vision of 'environmental collapse' comparable to the biological concept of trophic cascade (Kirsch 1997a). In addition to the immediate problems caused by the mine – the dead trees, the river dried up and full of mud, and the loss of fish, birds and other animals – the Yonggom perceive a variety of other changes in the world around them. They say that the sun has become hotter and burns their skin, that the rainy season lasts longer, that the stars and the moon are no longer as bright in the night sky, that rain harms the plants in their gardens and that the wind has become abrasive. They question the capacity of the world to sustain them. They use the term *moraron*, corrupted or corroded, like a piece of wood that has decayed or food that has become rotten, to describe the forests and rivers affected by the mine (Kirsch 1997a:149).

Yonggom fears about 'environmental collapse' reflect the new forms of risk created by the mine. In *Risk Society*, Ulrich Beck (1987: 154) analyses the social costs of environmental risks, arguing that modern forms of industrialisation have brought about a radical change in our relation to the world, in that our senses no longer provide us with an adequate understanding of our surroundings and the dangers posed to us by pollution.[13] Yonggom fears of 'environmental collapse' are directly related to their inability to assess the risks associated with pollution from the mine.[14] They lack a clear understanding of the way that pollutants may be transported in the air, through the water, or beneath the ground, and where the resulting lines of safety may be drawn.

We can say, as does Stephens (1995) for the Sami, that pollution brings about a 'doubling' of the world (see also Beck 1987:154).[15] Even if Yonggom sago palms show no outward signs of damage, they may not bear the normal quantities of starch. Even though their gardens initially appear prosperous, they may not yield the expected harvest. Surface appearances may mask the underlying reality of the mine's destructive impact. The 'doubling' of the world is of particular significance to the Yonggom given the association between productivity and the unseen world. Magic spells and *waruk* names, along with bird calls and other natural events, give the Yonggom access to this invisible realm. Pollution from the mine has also transformed this aspect of place, compromising the creative potential of the landscape. Conditions along the river prevent Yonggom magic from accessing the powers of the unseen world.

The other major change in Yonggom relationships to the world around them is that survival no longer depends on local resources. Today people are more likely to bring home food from a trade store than to parade game from the forest through the village. The largest source of income for the Yonggom is cash compensation for the mine's impact on their river and forest. These payments are scheduled to continue throughout the remaining decade of production at the mine. Instead of living off the land, the Yonggom must rely on compensation payments from the mine. In other words, there has been a shift from a subsistence economy dependent upon the use of natural resources to an economy based on resource rents, or payments made according to the value of their natural resources to others (Filer 1997b). Unlike ordinary resource rents, however, the Yonggom do not receive compensation in return for the consumption and use of their resources by others parties, but for the destruction of the productive capacity of their land as the indirect result of activity carried out elsewhere. Their environment is no longer a site of productivity, but a scene of loss. It no longer provides them with security, but confronts them with new, indecipherable risks.

New geographies of the imagination

In response to the challenges posed by the environmental impact of the Ok Tedi mine, the Yonggom have sought to perpetuate and expand their spatial understanding of the world. The location of the mine in the mountains to the north means that the character of the place in which the Yonggom live is largely determined by action taken at a distance. Rather than accept the terms brokered by the mine and the state, which permits the release of vast quantities of mine tailings and other waste material into their river system, the Yonggom embarked on a series of global journeys in which they successfully challenged the mine in a precedent-setting legal battle as well as in the court of public opinion (Kirsch 1996, 1997b). The result of their lawsuit was a negotiated settlement in June 1996 worth approximately US$500 million in compensation and commitments to future tailings containment.

By forging connections between these widely dispersed locations of capital and power, the Yonggom have been able to identify and respond to the space-time compression produced by the mine. Their campaign was successful because they were able to solicit support from an international network of environmental and legal activists, who helped them to regain control over their land. Andrew Strathern (1984) once used his informant's expression 'a line of power' as the title for a book examining the exchange cycles linking big-men across Melpa communities; new global 'lines of power' connect the Yonggom to their Melbourne lawyers and to political advocates around the world. Their new global imaginary extends from Port Moresby to Australia; in the Americas from Northwest Territory in Canada to New York, Washington, DC, and Rio de Janeiro; and in Europe from London to Amsterdam and Bonn, all stops along their campaign trail.

The Yonggom struggle against the mine is an example of so-called 'Lilliput strategies' of tying down and impeding transnational flows and globally dispersed work chains by linking 'local struggles with global support' and connecting 'local problems to global solutions' (Wilson and Dissanayake 1996:3). The international context of their activism follows the need to trace 'the complex and sometimes ironic political processes through which cultural forms are imposed, invented, reworked, and transformed,' as Akhil Gupta and James Ferguson (1997:4) have recently argued. Central to these processes is the effort to 'recover the concreteness of space [or place] that capitalism makes disappear' (Wilson and Dissanayake 1996:3; cf. Harvey 1996).

Yonggom magic-of-place and its ability to elicit new opportunities from the landscape might be compared to the mapping of global connections that manifests – as a form of productive activity itself – previously unknown links between places.[16] Their geographic imagination and their ability to map out alternative space-time and power relations enabled the Yonggom to conceptualise and realise an alternative future through the mechanism of protest and the formation of the global alliance that ultimately overcame Australia's largest corporation.

Future thoughts

Pollution from the Ok Tedi mine is not the only force that fuels Yonggom alienation from their environment. Urban migration, wage labour, Christianity and emerging national consciousness all contribute to the shift as well. These processes have parallels throughout Papua New Guinea, including areas unaffected by large-scale resource extraction projects.[17] Nonetheless, the mine's impact on the Yonggom has clearly pushed them further and faster in this direction.

These changes are reflected in Yonggom debates about the future. In 1996, I had several conversations in which people expressed their concern over the fate of the *aman dana*, the children of the future. Older people lament the fact that the youth of today have grown up without learning how to hunt and fish, to make string bags from tree bark, or to build houses without nails and sawn timber. The Yonggom have grown accustomed to eating rice and tinned fish and many young people resent the hard labour involved in gardening and making sago.

Their expectations for the future appear in two competing forms. Some of the members of an evangelical Christian church emphasise the importance of the coming millennium, which they expect to usher in a new era characterised by material prosperity. This future is defined in terms of accessibility to Western technology, which is sometimes referred to as the magic that enables Europeans to obtain whatever they wish simply by thinking of it. The Yonggom, in contrast, must work hard for their food by planting gardens, processing sago and by hunting and gathering. As Weiner (1995b:33) suggests, Melanesians may see 'our technology as a concealed or repressed form of *their* art [and magic].' Gell (1992b:58) refers to this as the 'effortless technology' that is a shadow of Euro-American production.

These European powers are also described as the magic of Digore, a mythological figure with the power over life and death who, like Jesus Christ, left after promising that he would eventually return. Digore is given credit for innovations such as houses with iron roofs and showers, stores full of food and other manifestations of Western wealth and technology. As one informant explained: 'When Digore returns … in the year 2000, there will be a new time and all of us will live together [as equals], European and Papua New Guinean.'

The millennial view holds that there is an opposition between the Papua New Guinea universe of villages, bush material houses and traditional technologies, and 'modern' life characterised by towns and technological achievements controlled by Euro-Americans. This perspective also takes the new emphasis on chronology to its logical conclusion, attributing agency to a particular moment in time, the year 2000. It is also a globalising discourse, synchronising the fate of the Yonggom to that of all Christians. It marks a conceptual shift as well: previously the Yonggom located the potential for transformation within the landscape, elicited by their magic or communicated by the other beings which inhabit the area or appear in their dreams. In contrast, this group attributes the power to bring about change to an abstract moment of time, which is by definition independent of place.

Other people in the villages have competing views on this subject even though they have comparable aspirations. A village catechist for the Catholic Church disputed millennial expectations popular among members of the rival evangelical church: 'The changes [they predict] are already taking place. The road to [the town of] Kiunga is coming closer and soon they will build a bridge over the Ok Tedi River and complete the road to the village. People have already begun to build permanent houses. Soon you won't see sago roofs at all, only tin roofs. Not long afterward, [electric] power will be coming in as well. These are the real changes and they are already taking place.' The catechist sees the technological markers of modernity – roads, permanent houses and electricity – as slowly diffusing across the landscape, moving steadily closer to the village. In keeping with Yonggom notions of creating possibilities by revealing the hidden potential of the landscape, the catechist suggests that their own political and legal efforts have brought these developments to the village.[18]

The two perspectives rely on opposing assumptions about the nature of change. The millennial scenario posits a succession of epochs that replace one another, a position familiar to anthropologists from an earlier generation of cargo cults (Lawrence 1964, McDowell 1988). In contrast to most cargo cults, however, the millennial transformation is not seen to be the result of human endeavours in ritual and exchange, but of chronological time with its synchronising frame and the universality that such a temporal grid implies. The resulting model of change is more conducive to conceptualising simultaneity with other people and places, a dimension of cargo cults that analysts have not always emphasised. The alternative perspective, represented here by the Catholic catechist, is that change is gradual, progressive and under way, but the direct result of human action and creativity.

The two views also incorporate very different perspectives on how the future manifests itself. The first scenario involves an 'effortless' episodic transformation in which the millennial future will manifest itself in the present at a pre-ordained moment of time, while the second scenario is dependent upon the productionist rhetoric of development and progress. Prior Yonggom understandings of the future as potentially manifest in the present correspond to the millennial view, but agency is displaced to magical forms of chronology, whereas the linear model of the future implies a productionist view that, while still grounded by place, involves forms of agency that are restricted to humans.

Conclusions

Before the mine, Yonggom ontologies emphasised place over time and the Yonggom saw their lives as unfolding in a landscape that they shared with other beings that possessed powers of agency comparable to their own. Historical relationships formulated in terms of the movement of people across the landscape have been interrupted by the destruction of the places that embody these memories. Yonggom concerns about these changes articulate with narratives in which the experience of loss is revealed through the landscape, shaping their response to environmental degradation. With the destruction of the landscape and the damage to resources along the Ok Tedi River, these relationships have been called into question and new risks have been created by the mine, raising the spectre of 'environmental collapse.' The Yonggom have turned to new forms of chronology to re-order their memories in a form that transcends place, leaving them estranged from the other beings with whom they previously shared the landscape. Magical forms of productivity have been jeopardised and the Yonggom face the challenge of a 'doubled' world that threatens the creative potential of place. The destruction of the landscape along the Ok Tedi River threatens to undermine their sense of place, history and enchanted forms of productivity, but also illustrates their capacity for overcoming these challenges.

This work raises an important question about the appropriateness of phenomenological realism for ethnographic description, which might be seen to exoticise the lifeworlds of the people under study. Yet Casey (1996) argues the connections between memory and place are a universal dimension of human experience, an assertion supported by the other contributors to Feld and Basso's (1996) *Senses of Place.*[19] While the experience of 'being-in-place' is valued by Euro-Americans, the Yonggom organise their most fundamental understandings of the world in these terms. Their ideas about enchanted forms of production emerge from their interactions with the living landscape, as does the importance they attribute to place over time.

Harvey (1996:315) argues that place-based phenomenological realism and alternative relations formed through capitalist space-time share a relationship of mutual encompassment. This is reminiscent of Weiner's argument that art and technology constitute a figure-ground reversal in which art reveals the relations obscured by technology (and vice versa):

> there is always a counter-invented world that emerges along with the intended objects of our conscious efforts but which remains concealed or unknown. This world is created as an unintended by-product of the focussedness of people's perceptions, and makes itself felt as a resistance to those efforts. It is not brought out directly, but only indirectly – hence our conventional terminology of production, construction, and ordering do not accurately characterise its origins. And because it is a reflexive effort, as it were, of intentionality ... it requires specific techniques, which are themselves non-productionist, non-relational, non-

constructivist, non-representational, to make them visible. [Weiner 1995b:42]

I suggest that this figure-ground relationship applies more broadly to phenomenological understandings of place and the technological and productionist view of the world established through capitalism. Not only does an awareness of 'being-in-place' oppose experiences produced through capitalism, it can also reveal its hidden assumptions, making it possible to conceptualise alternative modes of being in the world. Thus while Marx argued that capitalism works to replace space by time, it is not a matter of the establishment of irrevocable difference but the privileging of one perspective that partially obscures the other. The lesson is that differences between Euro-American discourses about environmental impact and how the Yonggom perceive the changes to their landscape need not remain a 'permanent feature of fragmented postmodern sensibilities' (Harvey 1996:285).[20] Working out these relationships is as important politically in terms of debates about the environment as it is theoretically (ibid.:285).[21] Perhaps the most significant element of the Yonggom case is the extent to which they have carried out the necessary interpretive work themselves in forging connections between their understandings of place and alternative modes of space-time relations imposed by capitalism.

In this ethnography of loss, I have emphasised the continuity of a tradition that puts place first in the imagination of the world. Feld and Basso (1996:9) have argued that 'place is the most fundamental form of embodied experience – the site of a powerful fusion of self, space and time.' Given the immense challenges posed by the transformations of their physical landscape, Yonggom use of chronological narratives, their creation of a global space for protest and their debates about the future might best be characterised as efforts to elicit new possibilities in their understanding of the dimensions of place. Like their hunting spells that work by revealing the unseen animals of the forest, these endeavours are intended to render visible their future along the Ok Tedi.

Acknowledgments

Research support was provided by a Royal Anthropological Institute/ Goldsmiths College Fellowship in Urgent Anthropology and the Institute for Research on Women and Gender at the University of Michigan. I am grateful to Roy Ellen, Steven Feld, Don Gardner, Eric Hirsch, Bruce Knauft, Andrew Lattas, Alan Rumsey, James Weiner and Michael Wood for their insightful comments on earlier drafts of this chapter, but any errors of fact or interpretation are the sole responsibility of the author.

Notes

1. In the Yonggom case, the consequences of environmental degradation outweigh the quiet but steady pressures of commodification that are the primary force of transformation elsewhere in Papua New Guinea (Foster 1995).
2. The economic costs of environmental impact include the consumption of resources with value as commodities and lost opportunity costs, although these assessments are limited by two factors: contemporary capitalism favours short-term calculation of economic costs and benefits, the problem that discourse about sustainability seeks to redress, and the assumption that economic valuation is universally applicable, which may 'bottom-line' values not appropriately reducible to economic variables or amenable to cost-benefit analysis (Rappaport 1993). The second mode of discourse is biological, measured in terms of impact on species biodiversity, population size and damage to particular habitats or ecological niches. The third mode of discourse is aesthetic, focusing on damage to the environment as something that may be experienced as pleasurable. Industrialised societies explicitly balance land used for production with landscapes set aside for consumption as 'nature' (Frykman and Löfgren 1987; Hirsch 1995:11).
3. Analysis of the conceptual consequences of environmental change may also facilitate comparisons of how ontologies of place and time change. In his book *A Place for Strangers: Towards a History of Australian Aboriginal Being*, Tony Swain (1993) adopts the extreme diffusionist position in claiming that the concept of time had no role in traditional Aboriginal beliefs, but was introduced through contact with other societies. In response to Swain, I would argue that place and time are not radically opposed, but mutually presupposing and that the processes that I have observed among the Yonggom should be understood as transformations of their ideas of place and time rather than the displacement of the one by the other. As Diane Austin-Broos (1996:6) wrote about the Aboriginal context, 'cosmology did not simply evolve as a different aspect to ontology emerged. Rather, large-scale change was accompanied by

extensive loss as well as transformation.' She also argued that the 'transition was more messy, more violent and more optimistic' than Swain indicates, 'in the sense of involving Aboriginal agency.'

4. Howard Morphy (1993:234-6) contrasts Aboriginal and European processes of locating and creating value in land: 'In the case of Europeans, history and landscape progress from outside to inside, history articulates with the landscape and releases its potential. Landscape is given a value by its place in history and by its economic potential.' In the Aboriginal case, however, landscape and myth (the Dreaming) are used to suppress or 'mask' the experience of history (ibid.:236, although see Rumsey 1994).

5. *Mimyop* is the Yonggom term meaning sorrow and loss. It is also the name of the heart, which is described as the source of this feeling. Mimyop is often referred to in terms of exchange; its archetypal form is in response to circumstances in which a relationship has been interrupted by separation or death. Sorrow is conceptualised in terms of a loss and the resolution of these feelings is contingent upon obtaining a replacement or compensation for what is missing.

6. Their influence is subject to the forces of entropy, however, so that the knowledge of these names must remain restricted if *waruk* names are to remain efficacious (see Evans-Pritchard 1929; Gewertz and Errington 1991).

7. Alfred Gell (1995:238) emphasises the auditory rather than visual dimensions of invisibility for the Umeda: 'an audible but invisible object was entirely 'present' in a way difficult for us to grasp, in that for us such an object is 'hidden,' however perceptible. The concept of 'hiding' in Umeda culture was, in fact, quite different from our own ... [implying] not invisibility, but the concealment of auditory clues, as in the silent approach of an assassin.'

8. Maurice Bloch (1995:66-67) describes the central value of 'clarity' to the Zafimaniry of Madagascar, which extends beyond visual domains. The name of their most powerful medicine, for example, means 'that which renders clear.'

9. Ingold (1996:145) argues that: 'Hunter-gatherers, in their practices, do not seek to transform the world; they seek revelation.' He (ibid.:144) challenges the dichotomy between 'peoples' practical-technical interaction with environmental resources in the context of subsistence activities, and their mytho-religious or cosmological construction of the environment in the context of ritual and ceremony,' arguing that their myths are not a 'metaphorical representation of the world, but a form of poetic involvement.'

10. Ingold (1996:131) makes a comparable claim that: 'Hunting itself comes to be regarded not as a technical manipulation of the natural world, but as a kind of interpersonal dialogue, integral to the total process of social life wherein both human and animal persons are constituted with their particular identities and purposes.'

11. Peter Gow (1995:49) suggests that for the Piro of the Western Amazon: 'the game animals of the forest and river [in contrast to garden crops] exist outside of human agency. Humans do not create them nor do they work to multiply them ... The game is produced as food by locating and catching it.'

12. In 1998, the Yonggom sponsored the largest *yawat* male initiation in many years, with the participation of more than fifty initiates (B. Nandun, pers. com. 1998). The ritual was promoted by the leaders of the protest movement against the Ok Tedi Mine, who were members of a single initiation cohort (*kaget won*) many years ago (Kirsch 1997b:126).

13. Beck (1992:19) argues that the defining feature of late modernity is the shift from a 'logic of wealth distribution in a society of scarcity to the logic of risk distribution.' By this he means that the central problematic of contemporary capitalism is not inequality resulting from the distribution of limited commodities, but how to manage the risks and hazards that result from the processes of production.

14. In contrast, Rolf Gerritsen and Martha MacIntyre (1991:36) argue that the "social, political and economic impacts of mining are principally about the distribution and redistribution of benefits,' largely discounting the problems of environmental impact associated with mining projects in Papua New Guinea. For an extended critique of their 'capital logic' model of mining and its discontents, see Kirsch (1997b:119-121).

15. Stephens (1995) describes how the fallout from the nuclear disaster at Chernobyl affected Sami reindeer herds, forcing them to depend on scientists and technology to assess the invisible dangers posed to them by radioactivity.

16. For example, several years ago I was contacted by a University student from a Yonggom village near the Ok Tedi River, who sought my assistance in mapping land boundaries for her lineage. She proposed using a GPS (Global Positioning System) unit, which establishes the absolute location on the Earth's surface through the triangulation of satellite signals. Like the sequencing of events by means of abstract, chronological time, this map would fix Yonggom places within a larger universal grid. It is difficult to imagine a better example of Yonggom attempts to place

themselves in the world system than this woman's plan to map Yonggom land using global positioning technology.

17. For an analysis of the concerns of rural Papua New Guineans in the absence of development, see Smith (1994).

18. Hirsch (1996) describes how the Fuyuge similarly envision change occurring as a result of their efforts to attract things from the periphery of their world.

19. For an extended discussion of the philosophical history of place, see Casey (1997).

20. Harvey (1996:287) argues that: 'It is precisely because of the transcending possibilities of such a global framing that we can register other forms of differences. Something may be lost in such a gesture – integrating into a hegemonic map of the world in order to demonstrate a particular cartography of domination and of power relations is no different than having to learn and use the oppressor's language in order to resist oppression. But something is also gained – the bringing to life of hitherto uncommunicated and hitherto incommunicable differences.'

21. This account is intended to challenge 'outright the absolutist perceptions and pretensions – the totalizing vision ... of the ahistorical treatment of space and time incorporated in conventional analysis ...' by examining how such relations are established (Harvey 1996:290). It calls for acknowledgment of the 'diversity in the social construction of space-time while insisting that different social processes may relate and that, therefore, the space-time orderings and the cartographies of resistance they produce are in some way or other also interrelated' (ibid.:290).

10

POISONING THE RAINBOW: MINING, POLLUTION AND INDIGENOUS COSMOLOGY IN FAR NORTH QUEENSLAND

Veronica Strang

Until the recent boom in tourism and development, the Cape York Peninsula in North Queensland was perceived as a 'remote area', heavily dependent upon the mining industry for much of its revenue. The region is particularly well endowed with minerals, having deposits of gold, silver, tin, copper, coal and lead. There are other forms of mining too – for example, bauxite mines at Weipa and marble quarries near Chilligoe. The peninsula contains equally diverse types of terrain, ranging from rainforested mountains along the eastern coast to the open savannah that borders the Gulf of Carpentaria. Its tropical ecosystems are rich in biodiversity, but also very fragile, with delicate aquatic ecosystems and friable soils which are easily eroded. Large river systems sweep down from the mountains and westwards across the plains, flooding rapidly in the wet season and dwindling to scattered chains of waterholes in the dry.

A number of Aboriginal communities are situated on the western coast, around the estuaries of these rivers. Via the waterways, which are central to their local economies, they are the inheritors of any pollution that takes place within the river catchments: soil erosion caused by overgrazing on the numerous cattle properties, and pollutants from the mines dotted around the river headwaters.

Aboriginal economies on the western peninsula remain partially dependent upon gathering and hunting, and particularly fishing. In the lagoons, rivers and estuaries there are about fifty-five different species of fish, some in great abundance. People fish regularly, using lines, drag nets, fish traps and spears. The Aboriginal communities also retain a cultural responsibility for taking care of their land. They are deeply concerned about the pollution of their water resources, and this has encouraged them to enter into a dialogue about land management with other land users on the Peninsula. This dialogue takes place within the wider complexities of state and national debates concerned with land use, tenure and title.

In this chapter, drawing on fieldwork with the various groups within the local community, I examine a dialogue between the Aboriginal people in Kowanyama, which is located near the estuary of the Mitchell River, and the other land users – including miners – represented in the Mitchell River Watershed Catchment Management Group. There are approximately 360 mining tenders in the Mitchell River watershed. Though Kowanyama's concerns about their activities are expediently framed in Western terms describing the impact of pollution on economic resources and biodiversity, the community's discourse remains, in reality, strongly focused on Aboriginal issues. It contains several important subtexts: about the maintenance of cosmological beliefs and constructions of nature, and about Kowanyama's relationship with the other land users in the catchment area and the wider Australian population. The elders' more obliquely articulated concerns about pollution and its effects upon the social and spiritual well-being of Aboriginal people suggest that, in many respects, the pollution of their water sources is a potent metaphor for the continuing colonisation of their land, the disturbance of their social and economic forms, and the 'cultural pollution' of their beliefs, values and identity.

Kowanyama

Kowanyama is a former mission community containing three different language groups, the Yir Yoront, the Kokobera and the Kunjen people,[1] who, prior to colonial settlement, inhabited land extending well beyond the current reserve area into the surrounding cattle properties. The mission was established at the beginning of the century to try to protect them from the incoming settlers. At that time Aboriginal groups faced a stark choice: they could come in to the mission 'to be saved', or they could stay on their own 'country'[2] by providing labour to the cattle stations in exchange for flour, tea, tobacco, and a measure of protection.

Before the invasion by Europeans, the Cape York Peninsula was densely populated, containing at least sixty different language groups (see Tindale 1974).[3] In common with Aboriginal people all over Australia, they shared a belief in a creative era, the Dreamtime, or 'Story Time' as it is known in North Queensland, in which ancestral beings emerged from the earth, formed all of the features of the landscape and then re-entered the land, to remain 'for all time'. In doing so they provided the totemic ancestors for Aboriginal clans, and collective land tenure acquired by clan membership and traditional systems of inheritance. Each totemic clan was thus defined by its particular ancestral tracks and story places which, according to Aboriginal Law, gave its members collective and inalienable title to that particular 'country'.[4] As Morphy has pointed out, Aboriginal systems of land tenure contain various flexibilities within an apparently 'fixed' ideology, allowing for clans, to 'merge and emerge' continually (1990:325). As Sutton and Rigsby (1982) have shown with reference to Cape York itself, this flexibility has become particularly important in adapting to the pressures of colonisation and its demographic disruptions.

In Kowanyama today Aboriginal social identity is still based on totemic associations with particular story places and ancestral tracks. Within their own clan country, or in a related tract, each person has a 'home' place, an *errk elampungk* from which their 'baby spirit' is said to have emerged. This is a Kunjen term composed of *errk* (place) *el* (eye) and *ampungk* (home) and may therefore be translated as 'the home place for your image'. Concepts of self are therefore embedded in place, providing each individual with her or his own location within a socio-spatial landscape.

As well as being fundamental to individual and collective identity, the land in precolonial times permitted a self-sufficient and sustainable gathering and hunting economy. The success of this economy was reliant upon a vast lexicon of bush lore: an intimate knowledge about local flora and fauna, the vagaries of the local climate, and a finely tuned system of land management. The latter included 'firestick farming' (Jones 1969): burning the country regularly to keep it clear and to provide green 'pick' for game animals.

The division of the year into extreme wet and dry seasons was reflected in the annual cycles of movement. There was much greater travel and gathering of different clans for social and economic purposes during the dry season when, for example, people would congregate from far and wide for communal fishing, using bundles of leaves to stun the fish in particular waterholes. Such gatherings also provided opportunities for rituals, corroborees, marriages and exchange. The water resources and aquatic life of the region were central to patterns of resource use,[5] and today fishing remains the most common – and some would say the most important – economic and social activity.

A spectrum of meaning

Water in its various forms runs through local cosmological beliefs, appearing in the majority of ancestral myths, and permeating every aspect of social and spiritual life. The ancestral myths abound with incidents at waterholes, creeks and rivers, and there are numerous water snake and flood stories. This centrality is reflected in people's bush names such as 'Floodwaters' (*Og-arrkal*), which refer to their ancestral connections. Story places are concentrated at waterholes, creeks, wells and other aquatic sites, and the actions of ancestral beings regularly created watercourses and lagoons. For example, at Emu Lagoon in Kunjen country, the Emu ancestor was dropped from a height during an airborne fight with the Brolga ancestor, creating an emu-shaped creek, for which the site is named (*Achamp ithan*):

> That's where that creek break out, that's where he dropped him. He wasn't creek there, not till that old bird bin come ... He never change, still the same. He had really Emu shape, that little creek. [Victor Highbury and Lefty Yam]

A map of the creeks, rivers and waterholes is also a socioeconomic narrative, representing kin relations and traditional exchange responsibilities within the community. Aboriginal discourse about the landscape therefore relies heavily upon water features to provide topographic, social, economic and spiritual orientation.

The waterholes, lagoons and rivers contain many serpent beings, and, in various guises, 'the rainbow' (*Ewarr*) [6]. The rainbow serpent is ubiquitous in the cosmologies of Aboriginal Australian groups, recurring in a variety of forms, and in every part of the country. As Merlan points out, it contains many complex meanings for Aboriginal people:

> Much that has been said about the northern 'rainbow serpent' complex focuses on its arcane and even secret and sacred associations with gleaming objects, with pearlshell, and even with semen (see, for example, Radcliffe-Brown 1930, Robinson 1956. Maddock 1970). These associations suggest themes of fertility and regeneration, and tend to be linked with a high culture of ceremony. The rainbow serpent is generally thought of as combining male and female properties. But in

> the Katherine area, the most widespread ideas of the rainbow serpent have to do with its aversion, as a native of place, to the foreign sweat and smells of people it does not recognize and its stormy response to incautious intrusions into its localities. [Merlan 1998:69]

In Kowanyama 'the rainbow' is presented in ways that are consistent with each of these interpretations: it is at once a sentient guardian, a source of knowledge, and the essence of human spiritual being and regeneration. In common with many cosmologies in the tropical north, it is believed to inhabit deep pools of water. Its presence is regarded as dangerous to intruders, and it must be treated with care even by those with whom it is familiar. Thus at *Errk ikow,* where the Mitchell and Alice rivers meet, a big snake lurks in a deep cave beneath the water:

> He will swallow you if you go in there ... you will never come out. [Maud May]

As in other parts of Australia, children and visitors have to be 'baptised' with water from the country so that it will 'know them' and its resident devils – of which there are many – will not harm them in any way.

> Have to baptise them little children you know ... or they might not be able to travel about, run around. Put them in water, put hand on their head, baptise them ... If he not baptised, well that devil might come along and grab that baby, take him away the bush. [Ronnie Smiler]

At Rainbow Story Place (*Og ewarr*), just north of the Alice River, a male rainbow serpent is said to have swallowed an ancestral being:

> He fishing in the waterhole there, rainbow come along and swallowed him. Finish him off, he be still there ... Cheeky one, you know, this rainbow a cheeky bugger. You leave him or he take you away for good ... That man rainbow, he kill you ... he never bring you back again same place ... If the man rainbow take you, you go for good. He eat you up! Yeah, finish, only bone left: he spew 'em out that bone. Take [you] like watersnake, same way ... swallow the lot. [Lefty Yam]

At Two Girl Story Place, between the Alice and Mitchell rivers, a giant ancestral 'fish' captured the two young women who tried to net it, dragging them along a creek and down into a lagoon:

> They try to stop that fish, but that not a fish, he bin rainbow ... They went after him, all the way they try and stop him ... but that fish still

> go underneath and lift them … Right down to the main hole … That's where he kill 'em. He drowned them two … He wasn't a fish, must be rainbow … The two of them go in there, either side the fish, and stay there. [Lefty Yam and Walter Parry]

The two girls were transformed into palm trees which may still be seen beside the lagoon:

> That's why them trees are there, on top there; they never come out again. Till today those two still there, them two young girls. [And the rainbow?] He stay – still there. [Lefty Yam]

At the main lagoon and at the original waterhole where the ancestral women began trying to net the rainbow, it is forbidden to disturb the soil around the water:

> We don't touch this place … We stir 'em up, bad for everybody. [Victor Highbury]

This is a common taboo at story places connected with reproduction – for example, a similar restriction applies at Baby Story Place further south. Only the owners of the country can safely conduct increase rituals at the sites, throwing handfuls of dirt in each direction in a ceremony to increase local resources. The throwing or movement of dust or dirt often features in such rituals, as does the use of water and the bark or leaves of particular trees. For example, at a waterhole along a story track formed by a 'poison snake', the Kunjen elders talk about a rain-making ritual which involved placing the leaves of a particular tree in the water:

> They make the rain here … They get the leaf and put 'em in the water down here. Big rain come, and strong wind, big strong wind, just like cyclone, he lift everybody up. If you walking about out here … lift you, carry you down to that main place over there, from this place. [Victor Highbury]

Taboos forbid the killing of snakes at the story place, and the ritual is connected with Darkness Story, in which the cyclone created by the ritual carries strangers trespassing into the country up onto a ridge where they are killed:

> They get that stick, a sword, a big sword, a wooden one, cut his neck here, kill him. Throw him over there in a big hole, ready to bury him, over there. [Lefty Yam]

In a reference which adds weight to Merlan's comments regarding the rainbow, Darkness Story also describes a gleaming ancestral shell which controls night and day:

> ... that shell, he let out the light ... Well he cover that night, see ... Old Tommy Koolatah, he showed me, 'there that shell now'. You can't go near him, he might shut'm off wind, you know, kill us ... he shut you in, he sort of close up, just like a mussel shell – sea shell. [Lefty Yam]

In its various forms and associations, the rainbow thus represents the protective sentience of the country, or what Merlan (1998:70) calls 'an autochthonous force that assaults the unfamiliar'. In response to ritual care, it provides resources for the traditional landowners and enables economic and social reproduction. It is also a transformative force, shifting ancestral beings and their human descendants from one dimension to another (see McConnel 1957, Myers 1986, Morphy 1990). At the beginning of life the rainbow sends human spiritual being outwards from the ancestral forces held in the land onto the physical plane of human life. In Kowanyama, spirit children are said to 'jump up' out of the waterholes into the bodies of women, and a sign is given, usually to the father or a close relative: this may be an oddity of some sort – an unusually large fish may be caught, or something may be a little out of place:

> Rainbow, well we all come from the rainbow. The whole lot of us ... The rainbow give us, show 'em to father or mother ... Only father see him. Baby here must be swimming round in the water here, playing. He pick him up, he go over there, sit down 'longside his missus. Well, that baby he must go back to mother. [Lefty Yam]

According to ancestral Law, at the end of life, part of the human soul goes 'west' or 'to the sky country' and a spiritual essence – the Dreaming part of the soul – must be ritually returned to her/his home place, the *errk elampungk*:

> Old people [the ancestors] used to send their spirit back to their own home, back to their own country. They all got different places, you know ... When I die they'll send my spirit back here ... They can't send us anywhere else. [Winston Gilbert]

Thus in the local cosmology, as in others (for example, Stanner 1966, Rose 1992), the rainbow is equated with the human soul. The emergence and reincorporation of spiritual being resonates with the many 'swallowing and regurgitation' myths associated with rainbow serpents which provide a powerful metaphor of death and rebirth in Aboriginal cosmology (for example, Eliade 1958). Roheim's (1945) Freudian analysis posits that the swallowing reunites humans with their mothers and allows for a 'rebirth' from a male being, thus sharing the responsibility for human reproduction. The rainbow is also implicated in many other transformative events during people's lifetimes; for example, in ceremonies enabling passage from one life stage to another. For instance, Hiatt (1984) connects the rainbow serpent with processes of parturition and separation which enable boys to be swallowed, hidden from female kin and regurgitated as men. The rainbow is also implicated in the acquisition of knowledge and authority. In Kowanyama, for example, those who want to become healers or 'witchdoctors' have to undergo a dangerous ritual which involves 'passing through the rainbow' to gain the necessary depth of spiritual knowledge.[7]

The rainbow is thus central to Aboriginal cosmology in Kowanyama and it represents, above all, the passage of time and change, and the flow of human social and spiritual being in and out of the waters of the country. As well as being the most critical economic resource, the rivers, lagoons and waterholes hold, in essence, the lifeblood of the community.

Red Dome Gold Mine

Owned by Niugini Mining, (which also has operations in Papua New Guinea) Red Dome is a large open-cut goldmine about 25 kilometres from Chilligoe. It has been running since 1986 and is technologically one of the most sophisticated mining operations in Australia. The ore is blasted, crushed and ground on site, and processed to various stages of refinement. The site therefore consists of a vast pit, a crushing circuit, a range of treatment plants and dams, and heaps of mine waste. A promotional video stresses that the company combines 'an efficient mining operation with sound environmental management'. Opening with a shot of sunlit water pouring like molten gold, the visual imagery places timeless shots of sunrises and wildlife alongside 'hard yakka' clips of men, machinery and dust, glamorised with semi-religious music and presenting an idealised picture of a more 'cooperative' relationship with the land.

Whatever the company's motives – and the miners at Red Dome admit openly that 'going green' gives them a strategic advantage, as well as saving money in the long run – their standpoint seems to have had some effect upon their values and their enthusiasms:

> Environmentalism ... it's become ingrained into the philosophy of the mine and the strategic planning ... it's not an add-on ... We do have an impact on the environment and we are going to try our very best to make sure it's as good as we can leave it ... I am sick and tired of the mining industry getting kicked like a dog ... but at the same time they are so keen on the money that is generated from us. And mining has played on that for a long time, the money that they make. Now it's time for mining to take the [environmental] lead and say 'we don't even need you guys to keep staring at us, we can do this'. [Bruce McCarthy]

Red Dome has followed through with the installation of innovative technology such as a cyanide recovery plant:[8]

> It has developed an Australian technology which has never been done before, so we're right on the cutting edge of it. And what comes out of here will probably become standard for mine sites and detoxification in the future ... Our aim is to leave the land in a stable, sustainable form and also to ensure that no contamination hazard exists here or migrating to outside areas. [Bruce McCarthy]

Red Dome expects to spend $1.7 million on decommissioning and rehabilitating the mine site when mining operations are completed. This will be enabled through seed collection and a nursery (which now produces 25,000 to 30,000 trees a year), massive landscaping and revegetation of the heaps of mine waste. The dam will become 'a sustainable aquatic ecosystem' with the introduction of flora and fauna from the nearby Walsh River – a 'recreational lake' for tourists and fishermen. The National Parks Department will inherit the nursery, and nearby pastoralists will benefit from the water bores and improved grazing. The mine plant itself will probably be dismantled, although Bruce McCarthy raises the question as to whether, as an example of leading-edge technology, it should not be preserved according to Queensland's recent legislation on 'Cultural Heritage':

> We're all very proud of our mine ... today's workings are tomorrow's cultural heritage. [Bruce McCarthy]

Red Dome plays a significant role in the economy of Far North Queensland, supporting, directly and indirectly, approximately 500 people. The miners see their activities as positive contributions, not only economically but also culturally. Like the pastoralists, they believe they are making 'proper' use of the land by making it productive, and they are very willing to engage in dialogues with other land users in the catchment area.

Dialogues in the Mitchell River Watershed

There are two major inputs into Kowanyama from the mining industry upriver. The first is a highly influential set of values which involves a particular vision of landscape, a material, Cartesian view of the land. Most modern miners have specialised training: they are chemists, bio-chemists, geologists or engineers. Their work invariably relates to a specific technical or scientific aspect of the mining process and they search for and extract non-renewable (and often invisible) resources that are fundamental to the land. For miners the environment is a set of material elements and 'prospects', and environmental problems are generally tackled with technological and transformational solutions.

The second and more obvious issue is the environmental threat posed by mining within the river catchment area. Whatever the positive gloss that companies such as Red Dome put on their operation, it remains the case that they are creating, in the headwaters of the Mitchell River, large tailing dams full of water tainted with cyanide and other chemicals. Though they try to be reassuring about their abilities to contain these poisons and prevent them from leaching or spilling into the aquifers and watercourses within the river catchment, the other land users in the region have a keen awareness of the environmental threat posed by such activities. They are also concerned about soil erosion caused by construction at the many mines in the region. According to a senior geologist, there is plenty of evidence to suggest that the waterways within the catchment are indeed being polluted by mining:

> The main problem is increased turbidity of the water due to the release of muddy waters from the washing plant. And the other serious problem is erosion of the areas disturbed by the mining and the numerous poorly constructed access and haul roads during the wet season. Because the stream beds have in the main not been rehabilitated, and been left as bare rocky watercourses devoid of their former

> trees, there's potential for extreme erosion in these areas and probably siltation of waterholes downstream. No-one has really studied in detail the effects of such mining in these tropical, largely ephemeral, streams. [Ian Withnall]

The Aboriginal people in Kowanyama have only recently gained access to this kind of information. In the past decade the massive intensification of land use in Cape York[9] has brought them into much closer interaction with other land users in the region, and necessitated a greater political engagement. Concern for the river is part of their overriding concern to regain control over their land and achieve a measure of economic (and thus political) self-sufficiency:

> We are trying to get our land back now ... We want to manage our country ourself. [Colin Lawrence and Thomas Bruce]

Like Aboriginal groups all over Australia the people at Kowanyama recognise that economic and political self-determination depends upon their ability not only to maintain ownership of their land, but also to have a strong measure of control over what happens on it. Even if they were to regain all of the land inhabited in precolonial times by the groups in Kowanyama, it is too late to re-establish a subsistence economy. There is a clear need for new sources of income, but employment opportunities in stock work or tourism are very limited. It is therefore unrealistic to oppose resource developments which may offer royalties, employment, or – possibly – financial or practical assistance in recompense for environmental degradation. As Connell and Howitt point out: 'the wealth generated by resource projects appears to offer a means of economic survival ... whilst expectations and needs have consistently increased...' (1991:4-5).

However, the Aboriginal community in Kowanyama is unequivocally protective of its relationship with the land. As one elder put it:

> It's better for us to ... stay right on the land ... If anyone comes along, some gold mining or something like that, comes out there and ask me can they do mining or something, well I won't quite agree to say yes. No, I'll say no, I'm looking back and saying this a legacy for the future. [Ezra Michael]

So Kowanyama's response to mining is enmeshed in wider issues of economic and cultural survival. The community is keenly aware that it has a limited amount of influence upon events – particularly those that occur 'upriver' – and believes that this is best realised through a constructive dialogue. Its response has been to initiate the Mitchell River Watershed Catchment Management Group (MRWCMG).[10]

> North Queensland Aborigines are taking the initiative to protect the Mitchell River Catchment, one of Australia's largest river systems ... Its catchment of 72,000 square kilometres is larger than Tasmania ... The primary aim of the group is to build up mutual trust and to bring about sensible management of water resources ... Some say it is a relatively untouched system, but warning signs have not gone unnoticed by the Aborigines. The elders plan to forestall the degradation that has occurred ... in other watersheds ... [*Conservation News*, December, 1990:7]

The MRWCMG contains a wide range of land and water users, but the major 'players' along with Kowanyama are the cattle station managers and owners; various conservation NGOs; Government departments such as National Parks and Wildlife and the Department of Environment and Heritage; and the mining industry. The discourse within the MRWCMG is dominated by the concerns of these groups.

> There is no question that the Western management paradigm is winning the discourse sweepstakes hands down. But a poignant dilemma remains: does the contemporary management mind have room for things that are important to indigenous people? [Cordell, 1991:6-7]

The exigencies of 'facing outwards' have had a number of effects upon the Aboriginal population. Social identity, previously located almost wholly in clan land and in the membership of particular language groups, has acquired a larger, more homogenous layer: 'the community' of Kowanyama. Thus there has been what Ernst (this vol.) calls a process of 'entification':

> Unity, one voice – three tribes of us in this community, so we all got to help [each other]. No good me talking about my country and forget about that country over there, no sense at all ... All one – pull as one person. [Thomas Bruce]

Part of this 'entification' involves the explication of cultural beliefs and values in terms more familiar to other groups in the area. For example, the complexities of language group and other affiliations are often subsumed in the more easily communicated concept of 'tribes'. In the process, social relationships and relations with land and knowledge about the local environment are to some extent reified and commodified as 'local culture' and 'expertise'.[11] Attempting to find common ground, the community's self-representation also contains a *bricolage* of the concepts of landscape presented to it by the other land users. Thus it includes the language of rural land management, tourism, conservation, and, to some degree, the deconstructive landscape of the mining industry.

Most of the other land users within the river catchment prioritise economic potentials, and in response the Aboriginal community stresses its own use of local resources.

> Every fish, we eat it that fish, we eat it. We don't waste it. We eat shark, we eat stingray: that our food ... In the early days my people used to look after the rivers because they knew where to get that good fish. [Colin Lawrence]

> These rivers are very important – we don't want too many white people coming in, taking fish from our rivers ... What they going to leave us with – nothing? [Ezra Michael]

Economic concerns, though easily communicated, mask a more complex set of issues related to the conflation of social and spiritual identity and its location within the landscape. There are strong feelings, not only about competition for the resources, but also about being invaded by too many people:

> We've got enough tourists coming in now, fishermen, all kind of men. [Harry Daphney]

Aboriginal landscapes contain a synthesis of ideas about land as the manifestation of ancestral 'bodies', and concepts of social and spiritual being that emerge from the land. As has been discussed elsewhere (for example, Munn 1986; Kolig 1987), mining is thus a particularly invasive disruption, epitomising more than any other form of development the invasion of Aboriginal land and self.

Merlan (1998) notes that in Katherine, rainbow serpents were presented as having been removed or killed by intrusive developments:

> ... they bin get'im little *bolung* [rainbow serpent] got'im dynamite take'im out. [p. 49]

> They killed that rainbow right there, the mother one, the big one. Killed it, pulled it out and took it away. I don't know what they did with it – maybe used it for oil. [p. 51]

In Kowanyama the location of ancestral beings in the water sources of the landscape, and their importance as wells of human life means interference with the waterways elicits similar responses. For example, an elder whose bush name is *Og im bon'ch* (moving water), talks about the need to prevent too many people from coming to his *errk elampungk,* a waterhole called White Water. If too many strangers visit this site, for which he has ritual responsibility, he believes that those connected with the place will sicken and die:

> People come up here this way, I got to put law against this place. Too many people come here, they make me go down too – make me sick. I got to stop them too ... Because this place I got story. Too much footprint on me ... I'm singing it here. Sometime they might sneak in here, we might get sick, old people go die then. [Nelson Brumby]

Invasion by strangers is seen to endanger the well-being of both the resources and the community, threatening the equilibrium of relations between them. This balance is seen as being maintained by customary land management – the ritual and physical care of the land. In accord with general precepts of Aboriginal land care, people in Kowanyama consider that 'proper' care of the land means maintaining a human presence within the country, and not allowing it to become 'wild' or 'dirty'. Keeping the country 'quiet' and 'clean' has moral, aesthetic and practical connotations (cf. Rose 1988). As well as clearing the country with fire on a regular basis, it is necessary to ensure that water resources remain clean and undisturbed. Thus an elder in Kowanyama explains how the freshwater wells in otherwise parched marine plains ought to be maintained:

> ... Water bin there all year round ... One old man must have bin put him there long time ... he used to get him water there ... He had to clean that water all a time, clean 'em, clean [out] that leaf, cover 'em up. Keep that water clean all a time. [Winston Gilbert]

> Same thing we talking about the rivers – we got to keep them clean, we got to keep the river clean, keep the lagoon clean. [Colin Lawrence]

In their dialogue with the miners in the river catchment, the elders stress their concern about the potential pollution of these resources:

> We don't know what's happened up the top end of the river, but all chemical off that mining ... [Kenny Jimmy]

> We worry about that mining company too, 'cause the mining you know, they can really muck em up the river ... I seen it, I've seen that big mine, looked at it. [Colin Lawrence]

People in Kowanyama are well aware of the aquatic ecosystems that carry pollutants downstream:

> We got a big mob river coming. We got the Palmer River, that comes into the Mitchell. We got the Alice River, that comes into the Mitchell. We got the Lynd, that comes into the Mitchell. We got the Walsh, that comes into the Mitchell. We see where that rivers come in, and they are mining up at the head of it. All kind of muck come down that river, just come into that one river, Mitchell River. And that Mitchell River feeding us ... Mine use too much water, muck up that water too. That's what we worry about, we worry about that thing, up this end here ... We want to keep that river clean for the future generations. [Colin Lawrence]

As well as borrowing rhetoric and concepts from scientists and conservationists, the emphasis on 'cleanness' signals a concept of order that is dependent upon Aboriginal control of the land, unadulterated by interference from anyone else. Discourse about pollution is linked symbolically with the anxieties about 'more people' coming in, and the need for Aboriginal control over these invasions of their land and lives. The spillage of chemical or soil into the rivers is equated with the disturbance of story places and all that they contain. Even a relatively small disturbance may be seen to have major effects; for example, where outsiders built a dam which threatened to spill muddy water over into a small waterhole at Cough Story Place, the 'father of the country' – that is, the senior owner – said that if this happened:

> ... and make it all creek – it be gone then ... We'll all get sick, everyone ... I'll get sick and die. [Winston Gilbert]

Pollution from upriver is seen as potentially disastrous not only to resources, but to 'everything':

> ... It will all be gone, finish. All the fish, all the animals, everything finish. [Kowanyama Council of Elders]

Conclusion

For the people of Kowanyama, pollution – whether human or chemical – of the waters in the Aboriginal landscape is, in essence, a poisoning of the rainbow – a flow of alien substance into the lifeblood of the community. It represents the disruption and oppression which outsiders have brought into this landscape, and the resultant injuries to Aboriginal well-being. The most recent 'pollution' from developments upriver is only a continuation of previous invasions by explorers, miners, settlers, missionaries, fishermen, Government officials, politicians, national parks employees, tourists, conservationists, (and, it must be said, anthropologists), all of whom have followed the watercourses westwards into Aboriginal land.

However much the people in Kowanyama try to frame their discourse in terms that fit into the paradigms of the other land users in the catchment area, beneath the expediently 'rationalised' arguments remain much deeper anxieties about protecting the rainbow: defending the land, and the Aboriginal Law within the land, from the pollution of outsiders.

> We want to try to keep our land because that land is very important for us ... White people, they think , 'oh this land' – they are only looking at the mineral, you know, oil and gold ... because they're so greedy, they want that land ... they don't worry about Aboriginal people. But we worrying about that custom bla we, see – our custom and our Law. White men – they got their custom too, but they're not like us. [Colin Lawrence]

Holding on to 'custom' requires a careful balancing of rather conflicting needs. Kowanyama has, on the one hand, to engage positively with the wider community, to translate Aboriginal concepts into 'land user friendly' terms, and to participate in negotiations about the control and management of the watershed. It often has made use of imported legal principles and management methods to establish its own claims to the land. Similarly, any hope of economic self-sufficiency is reliant upon the importation of European economic practices.

On the other hand, the community also wants to defend its ancestral Law from the 'pollution' of these different discourses and activities, and their deconstructive material landscapes, so that the Law retains coherence and provides a clear sense of social and cultural identity. This relies to some extent on a definition of Aboriginal cosmology that is in opposition to the beliefs and values of the other land users in the watershed and the wider Australian population. Aboriginal cosmology, with its exhortations to relive ancestral lives, is inherently conservative, and the land itself provides a firm anchor. In its own internal discourses, and in its dialogues with others, Kowanyama therefore has to both preserve and communicate its particular cosmology – to show the rainbow to the outside world, whilst also protecting it from pollution. Its battle to protect its waterways from being poisoned is, in essence, a battle to defend the Law.

Notes

1. There are other more subtle linguistic subdivisions within the community at Kowanyama, (see Alpher 1991, Sommer and Sommer 1967) but each is encompassed by one of the major groups.
2. 'Country' is the term used by Aboriginal people to describe their own tracts of land.
3. Tindale's maps show over 60 main language groups on the Cape York Peninsula, plus a cluster of 16 small groups on the east coast and 10 Wik 'tribes' north of the Mitchell River.
4. Details of the particular systems of land tenure and social organisation in Kowanyama are provided in Sharp 1937 and Strang 1997. Neighbouring groups are discussed in McConnel 1930, Von Sturmer 1978 and Sutton 1978.
5. Aquatic resources also form the basis of a local calendar, with seasonal terms such as 'fat fish time' and, 'crocodile egg time' (Strang 1997).
6. All of the language terms in this chapter are Kunjen, and there are several words for rainbow: *Ale* is the general Kunjen word, but in more formal, ritual language (Uw-Ilbmbanhdhiy) it is called *Ewarr.*
7. See also Taylor 1984 regarding similar beliefs in the Edward River community just north of Kowanyama.
8. This plant attempts to decontaminate the cyanide-tainted water used to leach the gold from the ore heap.
9. Land use in the region has intensified massively in the wake of the Cape York Peninsula Land Use Strategy, initiated in 1992 with nearly $9 million in Government funding for its first two years. CYPLUS, is intended to create:

> ... a framework for making decisions about how to use and manage the natural resources of Cape York Peninsula in ways that will be ecologically sustainable. [CYPLUS, 1993:1]

10. The MRWCMG is not an isolated case. Cooperative management of watersheds and other contained or interrelated ecological 'systems' has become a recognised necessity in many parts of the world. The Queensland Government has underlined its support for such endeavours with, in 1992, an allocation of $3.5 million for the Integrated Catchment Management Program. However, the management group in the Mitchell River Catchment is somewhat unusual in that it was initiated by the Aboriginal community.

11. As Jackson comments (in Connell and Howitt 1991:133), negotiations with outsiders, for example regarding mining royalties, have had this effect elsewhere, creating 'pressures to commodify and carefully define information and relationships in new ways'.

11

MINING, LAND CLAIMS AND THE NEGOTIATION OF INDIGENOUS INTERESTS: RESEARCH FROM THE QUEENSLAND GULF COUNTRY AND THE PILBARA REGION OF WESTERN AUSTRALIA

David Trigger and Michael Robinson

How might we characterise indigenous responses to large-scale mining projects in Australia? Certain aspects of negotiations with the wider society receive considerable public attention – in particular, the matters of protecting culturally significant land areas, environmental risks and monetary compensation (see, for example, Connell and Howitt 1991; Howitt et al. 1996). This chapter shifts the focus to a consideration of internal deliberations among Aboriginal people; we seek to investigate the social processes whereby mining developments are articulated with indigenous intellectual traditions about the significance of land.

In Queensland's Gulf country, Aboriginal participation during the past decade in various 'site clearance' surveys and especially negotiations over the Century Zinc Mine, have prompted both positive and negative local reactions to resource development projects. While some people have sought actively to lock into place a regime of potential benefits from mining, others have opposed it as inconsistent with cultural dimensions of Aboriginal relations to the land. Similarly, in Western Australia's Pilbara region, a series of new resource projects has been accompanied by large-scale and small-scale negotiations between government, developers and Aboriginal people. Indigenous communities who were largely bypassed during the major boom of the 1960s and 1970s now find themselves able to negotiate limited rights through native title and heritage protection legislation.

In both these regions of northern Australia, our research indicates that there has been at least as much conflict as consensus within indigenous communities, as people have encompassed new projects within their own understandings of nature and the landscape. Will creating a huge open-cut mine pit at Century interfere with subterranean spiritual forces? How is the permanently burning natural-gas flame on the Burrup Peninsula in the Pilbara linked to beliefs about the acquisition of fire from the sea by an important ancestral Dreaming? Such questions arise for Aboriginal people who are asked to 'clear' areas of land so that mining activities may be carried out. Furthermore, answers are arrived at in contexts where there are multiple pressures – from conflicting views within their own communities, as well as from industry and government parties who are likely to press for a prompt agreement that will enable the projects to proceed.

Sacred sites and the social production of meaningful landscapes

In the context of negotiations over resource projects, industry groups and governments commonly seek to solve the 'problem' of Aboriginal concerns about 'country' by identifying 'the sacred sites'; that is, the areas that can then be avoided in the development process (see, for example, Department of Aboriginal Sites 1994). Indeed, in both regions that are the subject of this chapter, there have been 'site clearance' surveys that might be regarded as having achieved some success in this respect. However, it has long been recognised among anthropologists that the significance of land for Aboriginal groups cannot be confined solely to particular bounded areas containing focal nodes of totemic meaning.

In this regard, the 'dynamic' nature of Aboriginal religious knowledge has been commented upon. For example, among Western Desert people, Tonkinson (1991) has described a 'characteristic openness' in traditional mythology (Tonkinson 1991:136), involving the production of 'new information' from spiritual revelations; this can include 'the finding of new [sacred] objects and the subsequent deduction of new links between sites and creative beings' (Tonkinson 1991:137). Also from a desert region, Myers (1986:64–6) recounts a particular illustrative case about how a distinctive topographic feature, not previously known as a significant site, was interpreted among Pintupi people in terms of familiar mythic details about the general country in which the place was located. This, he says, is geography operating as a code for producing meaning about the land; a 'deductive process … for explaining the existence of strange geological formations and shapes – or more generally why there is something at a place at all' (Myers 1986:66).

Merlan (1997:8) has coined the term 'epistemic openness' to designate this Aboriginal preparedness to interpret new meanings in the landscape; her own example from the Katherine area in the Northern Territory concerns the deduction of the significance of a singularly shaped stone found partly dislodged by grader work (cf. Merlan 1998, ch. 7). Merlan's discussion is focused especially on settings in which development projects are the subject of negotiations. Possibly the most heavily politicised context in which this issue has been salient in recent years is the controversy over developments at Hindmarsh Island in South Australia (see Bell 1998, Weiner 1999, Brunton 1996). For here there have been allegations of fraud in the production by certain Aboriginal people of particular meanings in the landscape. While discussing the possibilities of such fabrication in this case, Tonkinson (1997:11) has reminded us again of the 'considerable anthropological evidence indicating that the creation or revelation of new knowledge, including new sites, was intrinsic to Aboriginal religious life'.

It is this question of how Aboriginal people discuss country when it is the subject of contesting land use visions that we seek to examine in this chapter. Despite the ample evidence that, according to indigenous intellectual traditions, there is a certain quality of 'sacredness' associated with the entire landscape, and in fact that it can be difficult to distinguish 'sites' from surrounding areas (see Maddock 1991), confining the dimensions of land that should not be developed continues to be a major issue of conflict, both between Aboriginal groups and developers and among Aboriginal people themselves. What are the implications of what Merlan terms 'epistemic openness' for the process of negotiations over new mining enterprises? Does this notion help us understand how Aboriginal people conceive of transformations of the landscape by mining developments?

In the case of the Queensland Gulf Country, there has been a series of situations during the past decade where Aboriginal groups have found the task of making firm decisions about large-scale developments problematic. While this has been partly because of diverse indigenous views about possible benefits to be obtained from mining projects, it also follows from the nature of Aboriginal worldviews concerning the cultural significance of land. For it would seem that even the most definitive statement by particular knowledgeable senior people, that there are no 'sacred sites' in a particular area, is potentially subject to either direct challenge or what we would term circumspect 'worry' at a later date.

Certainly, it has been possible in the Gulf Country to survey various areas and establish from senior people that no major totemic (or 'Dreaming') sites will be affected by some of the proposed exploration activities such as drilling, road making and so on. However, especially if the area to be 'cleared' in this way is of substantial size, it is generally not feasible to inspect every aspect of the land involved. Thus, it is entirely possible that a topographic feature will be 'discovered' on some future occasion, and be interpreted as significant in one of a number of ways.

A striking example in the Gulf Country setting occurred towards the end of a survey where various sites were inspected and arrangements discussed as to protecting particular locations and allowing others to be destroyed by aspects of a large new mining development. This was an area that had also been flown over previously in a helicopter with senior Waanyi and Garrwa men to establish that no major totemic dreaming routes were present. However, while returning by vehicle through an area not visited on any earlier occasion, the survey party passed a particularly distinctive hill with rock outcrops protruding along its spine. People in Trigger's vehicle immediately discussed this hill with great interest, and the oldest authoritative man present proclaimed that it 'must be *ijan* [dreaming]', 'he dreaming that one' (that is, in his view, it surely must constitute an indication of a totemic force in the landscape). His comment is a good example of what Merlan characterises (1997:9) as an 'openness [of] epistemic attitude'; once stated in these terms by such a respected elder person, this is a matter that others would likely go on to discuss in the future, perhaps visiting the hill again if possible, and generally integrating interpretations about it into the regional Aboriginal worldview.

A case from the West Pilbara provides further illustration of this type of process. The Aboriginal people of the region[1] comment that land can be read like a book. Individuals commonly say that their language itself 'comes from the ground'. It originated from and is embedded in the landscape. To know country, then, one has to communicate with it, speak to the land and the flora and fauna in the appropriate language. In this context, the ubiquitous rock engravings of the region are likened to handwriting. They are symbols to be read in the same way, people suggest, as the Bible is to be read. Only those who understand the relevant language and have been through the appropriate ritual can interpret the landscape correctly.

In one survey, an Aboriginal field party encountered a stone arrangement which it believed might represent Warlu, the Water Serpent. This was not a previously known 'sacred site' but rather a feature discovered in the course of 'clearing' an area for further development. The survey party experienced some difficulty for it was made up of younger people who lacked confidence in assessing the site. Members of the field team felt certain that the arrangement was not due to chance and was not a 'natural' feature; their view was that the area needed to be examined further before development was allowed to proceed.

The issue was not so much a question of whether this was a newly recognised place of some significance but whether it was potentially dangerous in any way. Arrangements were made for a senior knowledgeable man from the community to visit the site and examine it for evidence of the presence of Warlu. He subsequently did this alone, insisting that others not approach, in case it was dangerous. After examining the site and talking to the land, he concurred with the survey party that it was indeed a Warlu place, but that this dreaming had departed because there had been so much development nearby; the site was within 100 metres of a major road. At the request of the developer, he marked out what he considered to be the site's boundaries and the company was asked to ensure that ground disturbance took place outside the area.

This was followed with a visit by a busload of people to show a broader community group how the survey had been performed. The bus stopped near the Warlu site and an outline of what had been discovered was explained to the party by members of the original survey team. During the visit a senior woman reported that she was having a vision in which she could see a serpent coiled in the air above the Warlu site. News of the vision quickly spread in the community when the party returned. It was regarded as further confirmation of the place as a Warlu site, but it also raised the question whether the Warlu had actually left the place as the senior man had earlier indicated. Several people who went on the bus trip complained later that they had experienced disturbed sleep patterns after their visit and attributed this to forces in the land which were warning them about the consequences of damaging the country. There was also community discussion about whether fencing the site, as the developer had proposed, would be sufficient to avoid enraging the Warlu. Finally, it was agreed that the fencing could proceed, provided senior men were present while it was being constructed.

Within the space of a week the Warlu site had become firmly established as a significant place in the Aboriginal community. People travelling along the road would point it out to others and recall the way in which the Warlu had revealed itself to the woman. During the fencing operation that followed, one of the Aboriginal men working on the project reported that he had inadvertently strayed into the zone marked out by the senior man and had been violently pushed to the ground by an invisible force. In an unrelated survey, the community objected to another mining company's plans to drill and blast on a nearby hill in case the action should enrage the Warlu. Although the senior man had stated that the Warlu had abandoned the site, the evidence of the vision, disturbed sleep patterns and the experience of the fencing contractor were sufficient to raise concerns that the site still possessed considerable totemic power. It was not the sort of place where risks should be taken.

Both the Gulf and Pilbara regions, then, encompass an indigenous cosmology that ultimately does not confine the intellectual framing of the significance of the landscape within a fixed body of knowledge with known finite dimensions. Certainly, there are specific, often named, sites with particular well known meanings; however, there is also a process whereby newly experienced objects, places and associated phenomena are integrated flexibly into indigenous worldviews, in a fashion consistent with long held traditional religious precepts. The two cases that we have presented concern making meaning out of newly encountered distinctive topographic features. We can also note how the 'power' of the inherent spirituality of the bush was apprehended through a vision and sleeping difficulties. These are examples of the ways a wide range of personal experiences are at times understood as prompted by the realm of spiritual forces – experiences as diverse as the witnessing of a surprising behaviour by an animal, an unusual shape of a tree or strange movements of wind or water.

These encounters with nature thus continue to be understood as 'saturated' with significant signs and meanings, as Stanner (1979:13) so elegantly put the matter, despite enormous cultural changes among Gulf Country and Pilbara groups over the past 100 years.

Furthermore, Aboriginal interpretations of country are commonly set within rich histories of occupation and use of the land, whereby sentimental attachments to the places people have lived and worked are mixed with nostalgia and respect for the fact that 'old people' once occupied the bush before the disruptions of European colonisation. 'Their hands touched these things!', exclaimed one person about her ancestors who used stone tool artefacts found lying near drill holes in the case of one survey in the Gulf Country. Her comment was made as others agreed that the stone tools must not be destroyed.

Given this type of social context, any particular decision about enabling a modification of 'country' for a large development, typically remains subject to an ongoing flexible pattern of interpretation implicit within Aboriginal cosmology and history. Indeed, a further Pilbara example illustrates how indigenous conceptions of spiritual potencies inhering in the land have been the basis for local explanations of the *very presence* of the actual valuable resources sought by developers.

Woodside Petroleum's natural-gas project was established in the early 1980s. The project involves the extraction of gas from the North West Shelf, some 130 kilometres north of the Pilbara coast, and its transportation by a sub-sea pipeline to a processing plant on the Burrup Peninsula. As a safety measure, the onshore facility includes a tall emergency flare tower that houses a permanent gas flame (see Plate 3). The flare tower can be seen for many kilometres and has become a landmark in the region. For most of the time the flare is visible from communities like Cheeditha (near Roebourne) as a glow in the western night sky. When there is low cloud over the coast, however, light from the flare can assume aurora-like characteristics as it reflects off the surfaces of the clouds. Aboriginal people at Cheeditha and Roebourne say that sometimes parallel shafts of light appear to pierce the sky from the vicinity of the flare, just like 'spears in the ground'. This is interpreted by some people as a sign that 'something' might be there; that is, something other than a simple safety device.

The Woodside facility is now a familiar part of the landscape of the West Pilbara and many Aboriginal people have visited the

Plate 3 (opposite): Flare tower, North West Shelf natural-gas project, Pilbara region, Western Australia. (Photo courtesy of Woodside Petroleum.)

project and understand the general processes by which the gas is extracted and refined. While the technological explanations may be accepted, however, the fundamental question of why there should be such a resource to exploit in the first place is a conundrum. A problem posed by the nature of the project is that, unlike the conventional mining projects with which most people are familiar, the natural-gas project does not have any observable mine site or, for that matter, any visible product. The raw minerals or geological composition of the terrain cannot then be inspected for clues about a possible cultural explanation for the substance. The source of the gas is also well out to sea and its processing takes place within a high security complex which is off limits to the public. The closest one gets to a visual manifestation of product is the emergency gas flare. With the same 'habit of mind' described by Myers (1986:67) for the Pintupi, the Aboriginal people of the area see in objects a sign of what might be there, and of what might explain the nature of things. The gas flare is seen as a sign of what might be behind, or what might explain, such a vast source of energy.

For example, we can consider the way one elderly woman related to Robinson that she often 'worried' about the natural-gas flare which she can see every night from her home. She was particularly puzzled by the occasional phenomenon of 'spears' in the sky and this had made her think that there might be powerful forces at work. She did not offer this as a definite conclusion, but tentatively put it that 'might be something there, that's what I think to myself'. She then went on to recall how her father, a *mabarn* ('clever man' or 'doctor') used to tell her that there was a very dangerous site out at sea where storms were generated. No-one except *mabarn* could visit it and 'drive' (that is, control) it to bring up rain from the sea. She did not know where the site was, exactly, but from her father's explanations it appeared to be out in the direction of the natural-gas field. This, she thought, might be the place that Woodside had discovered and the signs could be seen in the way the emergency flare behaved.

A widespread view in the community also links the flare to a myth about the origin of fire[2]. As the story is told, a family group possessed a single firestick which was stolen by Wagtail (Jirrijirri) Dreaming. Jirrijirri flew off over the ocean, somewhere north of Karratha, pursued by Karlamarna (Sparrow Hawk). As Jirrijirri was about to plunge into the sea with the firestick, it was plucked out of his hands by Karlamarna, who returned it to the shore where it was reclaimed by humans. Another named ancestral being then created songs about the event and engraved scenes from it into rock surfaces. Aboriginal people liken Woodside's emergency flare tower to the firestick (*thama*) and the flame appears as a constant reminder of the connection between the resource development project and indigenous understandings of the land.

Although the original myth said nothing about fields of gas or energy, people use it as a basis for explaining Woodside's find. The gas, they argue, could not have found its way to the bottom of the ocean by chance. The Jirrijirri story demonstrates, in their view, that Aborigines had knowledge of the existence of the gas fields long before non-Aboriginal technology discovered them. Aboriginal culture had contained the knowledge that a rich source of energy was in the sea, else why would Jirrijirri have flown out there with a firestick? The company's discovery is thus explained in terms of its technological ability rather than its knowledge of nature or the physical world of landscape and environment.

During recent discussions about whether compensation should be sought from the company, people argued that the source of gas is presaged in the Aboriginal belief system and is evidenced in narrative, song and engraving sites over a wide area. The company had only used its wealth and specialist equipment to extract the gas, whereas Aborigines had long known of its existence but lacked the technology to harvest it.

As has been evident during negotiations over mining in the Gulf Country of Queensland (Blowes and Trigger 1998; Trigger 1998), the indigenous view in the west Pilbara has thus been that natural resources are in an important sense owned by Aboriginal people; and it is expected that extraction of a resource will appropriately occur on the basis of adequate compensation. To this extent, indigenous conceptions of nature and the land have a direct political consequence for relationships with the broader society. However, negotiating over new development projects also presents substantial challenges for the management of social relations *within* Aboriginal communities. If designating the significance of landscapes is best characterised as a negotiated social process among Aboriginal people, it also is clearly subject to the ebb and flow of local politics. What are the internal political factors affecting indigenous constructions of meaningful landscapes in the 1990s?

Land, resource development and the politics of meaning

The case material presented in this chapter demonstrates the flexibility implicit within indigenous decision making. The significance of land is negotiated among Aboriginal people through interpretations of oral traditions and thereby through reference to extant knowledge of totemic geography. However, what is also clear is that there is a vibrant pattern of internal indigenous politics that determines such outcomes. Indeed, our data suggest that politicking and competition among indigenous groups commonly consumes more negotiating energies than does the process of dealing with industry and government parties. As Merlan (1997:9-10) puts it, the openness of epistemic attitude that she describes, 'is in contact with the basic materials of local politics'.

Recent literature has clarified the nature of Aboriginal politics that might be described as increasingly swirling around the operation of formally constituted indigenous corporations. Sullivan (1997:129) mentions that there is a lack of any effective internal political authority over disputing Aboriginal groups; furthermore, he depicts the constitution of indigenous land-holding groups as contextual rather than 'fixed in time and space', with memberships that vary 'for certain purposes at certain times' (Sullivan 1997:131). To the extent that this is so, can negotiations about mining (or other development projects involving land use) be based on dealing with indigenous representatives or spokespersons who remain stable in such positions over time?

In an important paper, Martin (1995) also characterises contemporary Aboriginal social organisation as resting on 'fluidity, negotiability and indeterminacy', with a 'stress on personal distinctiveness at virtually all levels of social practice' and a 'pervasive resistance to the imposition of authority' within the Aboriginal domain (Martin 1995:5). Thus, individuals who may hold prominent positions in indigenous organisations, are generally refused any recognised authority to represent the interests of constituents or members or to control the resources of the organisation (Martin 1995:8). Martin and Finlayson (1996) have developed this argument, describing the Aboriginal domain as typically highly factionalised, characterised by cross-cutting allegiances and thus entailing a strong tendency towards group fission and disaggregation rather than aggregation and corporateness (see also Sutton 1995). Individual and family interests constantly intrude, then, on attempts among Aboriginal people to achieve community-wide unity in negotiations with the wider society. Much energy goes typically into negotiating the internal social and political relationships that are of paramount concern in Aboriginal societies (Martin and Finlayson 1996:6).

These points hold considerable relevance for our discussion of indigenous responses to mining projects. In a social milieu where there is likely to be at least as much 'atomism' as 'collectivism' (Sutton 1995) how is it to be decided whether a particular area of land can be developed? Do designated individuals have sufficient authority to pronounce on the question of allowing country to be mined?

In the Queensland Gulf region, this has been a vexed problem during the 1990s for the multiple groups and residential communities facing decisions about the large new Century Zinc Mine. The pattern of decision making in relation to one particular area provides an illustrative example. In a 1994 study, a small cave site was found by archaeologists on a hill side just inside the perimeter of the proposed open-cut pit to be constructed as part of the Century project. The location of this site was previously unknown to Waanyi people; however, in April 1995 it was inspected with considerable interest by a party of some 26 individuals, with the outcome being that all present agreed it should be preserved. Stone artefacts were noted, as was charcoal on the surface of the cave floor. Similarly, on a hill nearby, a quite extensive stone quarry site was examined and the same decision was taken with respect to this site. The area was regarded as important as it indicated the historical occupation of the country by earlier generations of Aboriginal people. Comments included: 'that's part of our culture'; 'from our ancestors and all that, you know, from Dreamtime ... best to reckon it stay here'; 'our people been here, we don't know [how long], for many, many years, and we'd like to see this ... [hill] stay where it is, don't want it removed'.

However, over ensuing months, major divisions developed among Waanyi people as to whether it was best to sign a negotiated agreement that would enable the mining project to proceed (Trigger 1997). Disagreements derived from a number of dimensions of Aboriginal life, including diverse views about benefits offered and the chances of negotiating a better deal through applying pressure publicly to the company and the government; local level disputes unrelated to the mine also led to considerable argument between various senior persons as to just who held relevant traditional knowledge concerning the area. Both Aboriginal and non-Aboriginal people seemingly sought to exploit a tendency in indigenous social life whereby senior people joust forcefully with one another for reputations (Trigger 1992:111-118). Thus there developed a dispute, with considerable feeling and sentiment expressed from time to time, as to who among senior men and women could comment authoritatively on the matter of whether the stone artefact scatters and the cave site were too important culturally to be destroyed in return for what was being offered as compensation.

By late 1996, this split had developed quite bitterly among relevant senior Waanyi and Garrwa persons. One group of elder people, together with their younger supporters, pronounced that the site merely contained artefacts that could be found all over Waanyi country and that there was 'nothing there' in terms of spiritual significance. The other group (similarly diverse in terms of both genders and different age categories) remained vitally concerned that the area should remain 'as nature made him' and as it was when the 'old people walked around' (that is, when Aboriginal people still occupied the bush). Each side became locked into alliances with particular local and regional Aboriginal organisations and corporations.

Alliances were also formed with the non-Aboriginal parties. Both company and government personnel sought to support the senior indigenous people who appeared to be agreeing that the sites could be destroyed as part of the mine development. In December 1996, Queensland Government negotiators delivered a letter to those negotiating on behalf of native title groups; the lead Queensland negotiator reported on a meeting he had held with a widely respected senior man known throughout the region to hold expert status with respect to indigenous law. The Queensland personnel stated that they now had the agreement of this individual elder person that the cave site and artefact scatters could be destroyed; it was the Queensland negotiator's view that the senior law man believed the mine would benefit future generations of Aboriginal people through providing employment and associated opportunities.

The reporting of this alleged view of the law man occurred in a large negotiating meeting near the end of the right-to-negotiate period for Century Mine. It immediately prompted several other Waanyi men to complain forcefully that it was not solely the right of this one man to make such a decision. While his unsurpassed knowledge of 'law' was never challenged, his being positioned to speak for others on this particular matter was also never accepted. Yes, people acknowledged this man's word that no key dreaming sites were to be affected by the mine pit; however, others had different views about what was culturally valuable (in this case, a cave site and stone artefacts). Furthermore, there were some who repeatedly voiced their speculative concerns about whether digging down deep into the earth might disturb unknown spiritual forces; that is, potencies with no specific ('sacred site') surface level topographic manifestations but with a general presence underground: 'They gonna wake up that Rainbow [Snake], I'm sure', as one man put it.

Thus, in this highly contentious case, there was a complex mix of internal indigenous politics and strategies pursued by the industry and government parties, which influenced the way meanings in the landscape were negotiated among Aboriginal people. The tendency towards an open epistemology was harnessed to various causes and alliances. No clear decision about whether the cave site and artefact scatters should be preserved or destroyed was forthcoming by the end of the negotiation period. For there emerged no process of indigenous decision making that could be accepted as authoritative among all groups and individuals.

In the case of a similar dispute in the Pilbara, we can describe what appears to be a more consensual outcome. Here a community was faced during the early 1990s with a respected senior man's assertion that a particular hill was a significant Dreaming site. A mining company had arranged an inspection of its proposed area of operation, including a prominent hill which was to be the source of its ore. The hill itself had been mined over many years by several other mining companies which led the proponent to anticipate that there would be no objection to its continuing to develop the area. Not only had the hill been substantially mined, but there had been two separate site surveys in the past which had reported that it did not have any cultural significance. The company had only arranged a further ground inspection because of an archaeological survey which had identified a previously unrecorded site on the plain below the hill. It wanted to establish whether the Aboriginal community would object to this archaeological site being disturbed, and included an inspection of the hill to familiarise the community with its plans.

After driving to the summit of the hill, a senior man in the party declared that this high ground was the metamorphosed body of an important Dreaming figure. He named it and pointed to other features in the surrounding landscape which he said were associated with the being and its movements. On the basis of this assertion, a further site survey was then initiated to establish whether there was community support for this particular man's assertions and to discover why the earlier surveys had not reported similarly on the hill's significance.

During the further survey, the senior man maintained his position that the hill was culturally significant. He said that he had been told about the area by two brothers, now deceased, who were regarded by the community as the traditional owners of the area. The daughter of one of the brothers lent some support to his views by saying that she had heard that such a hill was located in the general area, but no-one else claimed to have independent knowledge of such a site or could corroborate the senior man's explanation.

This view propounded by such an influential individual created political and economic difficulties for the community. Negotiations had already begun with the mining company concerned, which had indicated a willingness to pay compensation generally for its use of the land and to employ community members on the project. If the ore body could not be mined, however, there would be no project and therefore no compensation or jobs. The seniority of the man making the assertions of significance meant, however, that there was a potential for political repercussions if people opposed his point of view or contradicted his statements. There were also implications for those who had participated in the earlier surveys and indicated that the hill was not significant. It emerged that the senior man had not been consulted during those earlier surveys.

Although no-one in the community came forward to verify directly what the senior man had said, it was also the case that nobody was prepared to contradict him publicly. Those who participated in the earlier surveys largely withdrew from direct participation in the resolution of the problem. They neither accompanied the new survey nor took part in subsequent meetings about it. In private one of the leading members of the previous surveys stated that he would accept what the senior man said about the future of the mine and would not openly oppose him.

After a period of some weeks, while internal discussions took place about the advantages and disadvantages of the project, the senior man resolved the dilemma by declaring that, in his opinion, the hill had been so badly damaged by previous mining that any objection to further development would be pointless. While by no means changing his view that the hill was significant spiritually, he announced that he would not object to mining proceeding at the site, and the broader community once again was able to turn its attention to negotiating with the company for 'compensation'.

Conclusion: The politics of culture and the articulation of indigenous interests in mining negotiations

In this chapter, we have sought to depict two key features of Australian Aboriginal responses to large-scale mining projects. Both concern deliberations occurring among and within Aboriginal groups in the broader context of intense negotiations with industry and government.

First, we have discussed how the significance of land itself must be resolved through processes of interpretation and often argumentation among relevant Aboriginal people. This is not a matter of simply consulting a fixed body of knowledge about particular 'sacred sites', but rather involves prominent individuals interpreting signs in the landscape according to a broad set of beliefs about the general spiritual forces underlying the world of immediately observable topographic and other phenomena. Furthermore, the signs taken into account can include the very existence of discovered valuable resources that are sought by the institutions of the wider Australian society.

This is a process of 'reading' the landscape according to what we might term a set of rules (or what Myers [1986:66] calls a 'code') that derive from bodies of customary 'law' and traditions. While there are important elements of personal and collective interpretation at the core of such 'readings' of the country, and thus ways in which the process is indeed appropriately regarded as 'epistemically open' (Merlan 1997), it is also important to be clear about the constraints upon this intellectual practice. Certainly, it would be inaccurate to suppose that the cases we present in this chapter simply involve 'inventiveness' or 'fabrication' on the part of ambitious individuals. Our data suggest that features of the landscape, including rock art, engravings, old camp sites and a wide range of environmental phenomena, will be typically understood according to intellectual principles embedded deeply in regional cultural traditions. We might best conceive these principles as constituting a fundamental level of underlying regional indigenous customary 'laws' (Sutton 1996); that is, assumptions and precepts which arguably remain robust through much cultural change, and which continue to play a critical role in the way Aboriginal communities work out rights and interests in land at the local level. Decisions about the appropriateness or otherwise of large-scale mining developments are, to a considerable extent, likely to be deliberated upon among Aboriginal people in terms of interpretations about country that are driven by such broad customary knowledge.

The further issue in this chapter is the political nature of the interpretive process among the Aboriginal communities with whom we have worked. This is to demonstrate that there is a vibrant indigenous body politic in operation which must be understood if we are to depict accurately the way in which Aboriginal people respond to large-scale new mining enterprises. While regional customary 'law' provides the context in which indigenous views about mining are negotiated in the Gulf Country and the west Pilbara, patterns of local indigenous politics can involve considerable conflict about appropriate uses of the land. The brief case materials given here indicate that a competitive politics of reputation, especially among senior individuals, will commonly operate in a relationship of some tension with the more collectivist imperative to draw on shared customary knowledge in establishing the meaning of land and the appropriateness of its 'development'.

Finally, such competition over interpretations of cultural knowledge is commonly situated amidst other patterns of social relations whereby different indigenous corporations, families and individuals must vie for access to financial resources from a host of government and other sources. There is typically a mix of cultural politics and material aspirations that constitutes the setting in which indigenous interests are articulated in the context of new resource development projects. In our view, understanding indigenous responses to such projects, thus requires a sophisticated recognition of the resilience of cultural beliefs about land, while also facing squarely the implications of local politics driven by the material realities in people's lives.

Notes

1. References to the West Pilbara region in this chapter draw predominantly from the Aboriginal communities of Roebourne, Cheeditha, Wickham and Karratha. These communities consist in the main of Yindjibarndi and Ngarluma peoples but also contain significant groupings of Kariyarra, Kurrama, Martuthunira and Banyjima (Wordick 1982; Edmunds 1989).
2. See Brandenstein (1970:278) and Wordick (1982:257) for separate accounts of the myth.

12

DEVELOPMENT, RATIONALISATION AND SACRED SITES: COMPARATIVE PERSPECTIVES ON PAPUA NEW GUINEA AND AUSTRALIA

Francesca Merlan

This chapter compares some aspects of indigenous response to resource development in Papua New Guinea and in Aboriginal Australia and the role of the state in both areas, focusing on discussion of the recent Australian scene. The focus of the comparison is the nature of cultural and cosmological change and continuity in both situations.

The wider conditions for change and continuity are rather different in the two situations. In Papua New Guinea, an approximate 97 per cent of land area continues to be held under customary tenure. The country is relatively resource-rich, and its minerals, timber and other resources are much sought after. In many instances, New Guineans have been initially positive about 'development', a word they have enthusiastically taken up since the times of the *kiaps* and the project officers who first familiarised many of them with it (see Jorgensen, this vol., Wardlow, this vol.; cf. Sahlins 1993). The state is relatively under-developed and has limited local penetration; hence resource developers, to get projects moving, may quickly come to play a role at local and regional levels (see Sagir, this vol.). They do so by taking on some of the organising and structuring functions that a more developed state might otherwise assume. Given the often quite direct interface between resource developer and local communities, the former have had to become responsive to local concerns, forms of organisation, and interaction, especially in the recent period. Especially following the unrest that surfaced with particular intensity since 1989 around the large copper mine on Bougainville, resource developers have seen the Papua New Guinean situation as potentially volatile. They perceive that initial local acceptance of large-scale development projects may not last (Kirsch, this vol.), partly because the state cannot guarantee close management of on-the-ground communities.

In Australia, the situation has been very different and, I argue, is currently changing. In most parts of the country, Aboriginal people were effectively dispossessed of their land and of control over it from an early period of colonial settlement. Only recently, over the past three decades or so, has a more fully fledged, modernising, postwar state accepted major responsibilities for Aboriginal affairs, guaranteeing some central oversight of what had been largely managed by the various states and territories of the Australian Commonwealth. With the assumption of responsibilities by the Commonwealth has coincided a tendency towards the re-valorisation of cultural difference, a trend found more widely in the liberal, Anglo-American world. Aboriginal people were earlier expected to 'assimilate', and become like members of a notional Australian 'mainstream'. From around 1970, however, the Commonwealth promulgated a much more liberalised definition of who might be an 'Aborigine', and a policy of 'self-determination', both granting considerably greater latitude to Aboriginal people to be 'themselves', rather than be moulded as Euro-Australians. Since that period, there has also been considerable land rights activism. This resulted (from 1976) in the making over of large areas of land to Aboriginal inalienable title in the Northern Territory under a land claims process instituted by federal statute in that jurisdiction (the *Aboriginal (Northern Territory) Land Rights Act 1976*). More recently, the land rights movement has culminated in the High Court case known as Mabo (1992), in which native title was found to persist in the Torres Strait. This paved the way for enactment of the *Native Title Act* (1993), which lays out a regime under which indigenous people (as well as others) may make application for determinations of the survival of native title elsewhere on the continent, and sets up a regime for negotiation and arbitration of interests between native title claimants, or proven holders, and others.

In the light of this history, one of my arguments concerns the changing character of episodes that, from time to time, have emerged in the Australian public domain as disputes over Aboriginal relationships to places, or sites, and have come to be known as 'sacred sites disputes'. Over a twenty- to thirty-year span, and especially as particular disputes outran their local and regional dimensions and came to occupy national attention, they were often developed as what I will call 'full culture' arguments. That is, what was identified publicly as an Aboriginal position in these disputes typically involved rejection of resource development or some project around which the issue of relationship to the site had developed in the first place. This 'Aboriginal position' was frequently argued on the basis that disruption of development of particular locales should be rejected because of the far-reaching significance of disturbance of them. Such disturbance was also said to introduce a deformation into a more inclusive indigenous system of ideas, representations, perspectives on the world, values, understood to be attached to and shared by some bounded social group (however poorly known and represented to the wider Australian public). In short, it would disrupt a 'culture' in the inclusive and systemic sense of this term from its origins in 19th-century Germany, and through some of its history as a key term in 20th century (especially North American) anthropology.

As these disputes proceed further and further into the national domain, there emerges an emphasis upon the 'symbolic' significance of places to indigenous people, as distinct from and counter to the material interests in them as expressed by resource developers and others. Impelled by the necessity of the parties to shape public opinion, there results an image of indigenous people as more oriented towards issues of culture and meaning, and by contrast, the representatives of dominant interests as more oriented towards issues of profit and technical mastery. This contrast however obscures the anthropological recognition that no distinction between the material and the symbolic is typically made in this way by indigenous people themselves. The contrast emerges when they are brought as participants in these disputes into the wider public arena of debate in which their own ways of putting arguments are thoroughly re-oriented to dominant schemes of meaning and practice. These schemes typically presuppose a divide between the material realm and the symbolic. In these terms it is also generally further assumed that, through development, material value would accrue to the wider community, including Aborigines, while non-development would be a value principally, if not exclusively, for the Aboriginal people concerned.

For present purposes, the conditions for making the 'full culture' argument can be located in the increased readiness after World War II to recognise a culturally differentiated world, in rising skepticism about the universality of (Western) values of progress, and in reaction to intensifying processes of what is now called globalisation. In this last context, globalisation is viewed as countering the avowed values of cultural difference and multiculturalism.

In this chapter, I want to suggest on the one hand that the social conditions for making the 'full culture' argument are changing and have changed. On the other, I suggest that some of the institutional processes and changes currently going on in Australia, such as the formulation of Native Title, are being accompanied by a decrease in acceptance of cultural differences as untranslatable and incomparable. There are parallels between the redefinition, 'downsizing' and lessened systematicity of notions of 'culture' in these public domain issues and institutions, and in academic disciplines concerned with culture and related notions.

With respect to the empirical details of sacred sites disputes, I suggest that there is a movement from the public constitution of indigenous positions as ones of non-negotiability, or radical difference, towards other positions constituted as negotiable, or at least as drawn within a framework of negotiation and a framework of compensation – where this often, more bluntly put, is assumed to mean 'price'. Previous tendencies towards identification of Aboriginal 'special status' are shifting towards identification of Aborigines as having an interest, and still perhaps based on an originary or indigenous identity, but nevertheless as something more like 'co-stakeholders' in a variety of processes that, increasingly, recognise their contemporaneity and diversity, rather than pastness. Underlying this shift from non-negotiability to negotiability, is what might be called a quasi-intentional 'strategy' of the nation-state, or a way of organising institutional change to detach itself from previous entrenched forms of impasse between Australia and its indigenous people. This is a move towards assertions of comparability between institutions of the dominant society and those of indigenous society, now recognised and revalued. Such assertions of comparability have occurred, for instance, in the finding which underpins both Mabo and the subsequent *Native Title Act* that 'Native Title' is justiciable at the common law. I think that in the recent history of this shift from non-negotiability of cultural difference to a frame of negotiation, there are reasons why, over the past few years, events around a major sacred sites incident at Hindmarsh Island, in South Australia, have been seen as a watershed. I will expand on this below.

The Papua New Guinean scene differs greatly from Australia, and indeed from Fourth World contexts in general, in that Aboriginal people in Australia are held accountable to hegemonic notions of authentic, relatively fixed cultural process as at the basis of claims they can make about the significance of sites. As I will further discuss, they have been held to accounts of the significance of places that come from the past, and demonstrably pre-date any development or other proposal. This insistence upon cultural stability, if not inertness, has been a factor in the way disputes over resource development projects and sacred sites have emerged in the public domain in Australia. In contrast to Papua New Guinea, there has been a recent history of considerable Aboriginal resistance to resource development projects – not surprising in terms of the history of disempowerment, expropriation and marginalisation. But I think one has to understand this at least to some extent in light of the nature of the historical relationships between indigenous people and settler society, rather than in terms of some deep-seated cultural difference between Aborigines and New Guineans in this respect. And it is indisputable, I think, that what is represented at the national level as the nature of indigenous positions is the product of very complex inter-cultural and power-laden construction – not straightforwardly authored at the local level and directly transmitted to the national public.

The liberalisation of Aboriginal affairs policy is conventionally dated back to the Whitlam Labor government. 'Assimilation' was discarded as Aboriginal policy, and 'self-determination' was proclaimed. There was rising public recognition that Aborigines had been treated oppressively, and a determination they should be treated better in future, or at least, that outright oppression should cease and some sort of 'fair go' extended to everyone. Though the legislative and regulatory 'fair go' does not erase historical inequality, it can be an improvement over the demand that everyone be the same.

Along with this went an argument that radical difference might be recognised, that the 'self' of 'self-determination' might be very different, and not brought within the same frameworks as other Australians. As mentioned, the *Aboriginal (Northern Territory) Land Rights Act 1976* was brought into being as a special beneficial statute to recognise Aboriginal rights to land. The enactment of this special statute was a direct outcome of the decision in the now-famous Gove case of 1971 (Williams 1986). In that case, Aborigines of Yirrkala in north-eastern Arnhem Land brought suit against the Commonwealth and Nabalco, a largely Swiss-owned mining company, for the excision of a mining reserve from the Arnhem Reserve to develop the large bauxite mine that in fact now exists in the region. The Aborigines brought suit, claiming their rights to land had not been recognised. The outcome of a long hearing was the decision by Justice Blackburn that, though certainly a complex system of land tenure had been placed in evidence before him, it did not amount to possessory right at the common law, lacking features of exclusive possession and right to alienate that he held to be elements of ownership at the common law. The *Aboriginal Land Rights Act 1976* was thus formulated as a special beneficial statute designed to recognise the special character of Aboriginal land tenure, since it apparently could not be recognised at common law. This is a measure of growing willingness to look at Aboriginal affairs as involving a distinct 'culture', or 'cultures'. It continues to involve a notion of bounded difference that assimilation policy also had presupposed, but had treated as something to be superseded.

During this period and for some two decades afterwards, I have suggested, there was some willingness to recognise claims by Aborigines as involving 'total cultural claims'. Most of the episodes that made their way into the public domain involved resource development proposals, and many of these took on a quite oppositional character, with Aborigines and/or Aboriginal interests represented as opposed to development. The most notable of these episodes concerned 'sacred sites', in which the claim was made that development could not take place because of the character of relationship that Aborigines had with the place concerned, and because of the significance they considered to inhere in it, a significance even more strongly represented in immaterial, symbolic and religious terms.

I have suggested that many of these disputes, as they emerged into the national domain, came to take on a total cultural character. This is not to suggest, however, that there was unanimity of opinion about development projects within the local communities which had primary associations with the locations in question. In asserting that there were differing views, I do not wish to deny the genuineness of Aboriginal concerns about resource development. These include issues of the specialness of places and the danger of tampering with the country as it is given and known to them, environmental matters as these were conceptualised, and a general desire for recognition and social justice. However, the situations were complex, and answers to the types of questions that were being posed often did not appear to be given by tradition. Distribution of information and levels of understanding were very variable at local levels – there was very limited consultation within the community. By contrast, there were higher levels of information within Aboriginal bureaucracies, but also some tendency to support oppositional positions within certain key ones. Finally, there was limited incentive for local Aboriginal participation because of the extremely weak, not to say marginal or excluded, position that Aboriginal people tended to have with respect to any development proposals. A variety of agents were in play, including state and federal agencies, and resource companies, who often had direct access to Aboriginal communities and individuals for periods of time, and thus could foster the kind of personal ties that are important for securing acceptance. Aborigines were seldom consulted fully or meaningfully, and consultation is difficult given its power-laden nature.

In some cases, Aborigines were being asked new questions about what might or might not be allowable at given places. In others' eyes, they had not had any authority to do so, before the passage of such legislation as the *Sacred Sites Protection Act* in the Northern Territory, state legislation complementary to the federal *Land Rights Act.*

In any event, what can be meant by 'tradition' is always relative to some present. Contexts that are newly articulated in this way tend to permeate each other (Wagner 1975:58). Recontextualisation of past experience does not leave it unchanged. To give relationship to place a new articulation in the present is inevitably to re-shape some of its terms and importance, an issue raised in the discussion of some of the cases below.

Given the conditions briefly described, it can be understood why the working out of an Aboriginal position in the national arena often has tended to involve a 'full culture' argument. Given that Aborigines were at best marginally included in consultations, the strongest Aboriginal position about what might happen was shaped as part of an inclusive world-view. This world-view was often described as non-negotiable in terms of Aboriginal culture, and as outside the framework of material compensation. More particularly, the relevant relationships of Aborigines to places were frequently represented as part of that aspect of culture concerning which the wider public is most respectful: religion. The message had gotten through, at least to sympathetic ears, that 'premodern' cultures are suffused with religiosity, so familiar institutional divisions did not apply, and religion could not be bounded off from other domains. What might be decided about a development project was not only religiously informed, but inseparable from other aspects of culture. Further, religion was not typically seen as subject to practitioners' choice or disposition, as it might be in Western social locations, but rather as defining the very ground from which any choice is possible, and thus as essential to identity (see Weiner 1999; n.d.).

Sacred sites disputes

Let me introduce and then consider a few 'sacred sites disputes' that have risen to national prominence over the past twenty years or so in Australia, largely in the context of development proposals of one kind or another.

Noonkanbah

'Noonkanbah', which existed as an aggravated sacred site dispute from around 1978 to the end of 1980, might now be seen as a classic case. Aboriginal people had recently moved back onto this station in the Kimberley (see Map 1, p. ix), and established a community, after a period of physical alienation from it. Noonkanbah had been developed from 1886 by the Emanuel family, as the first permanent white presence in this part of the Fitzroy River basin. Aborigines had become labour in the pastoral industry, and remained in numbers on the station, under hard management. By 1971, following the introduction of award wages, the last Aborigines left and subsequently spent a troubled five-year period of residence as fringe dwellers around the town of Fitzroy Crossing.

In 1976 the Commonwealth Government, and specifically the Aboriginal Land Fund Commission, purchased Noonkanbah for what became incorporated as the Yungngora Community, and people returned. By 1978 a hectic race was on to claim ground in the Kimberley oil and diamond exploration boom in this area where mining had heretofore not played a great role. By May 1978, 497 mineral claims had been pegged on Noonkanbah, by about 30 different mining interests. Aborigines were alarmed by this level of presence. Conflict intensified particularly with the development of oil exploration by AMAX Iron Ore Corporation, whose parent company was the largely American-owned Amax Incorporated. A prospective site had been located at Pea Hill, or Umpampurru, a relatively prominent feature in the landscape and thus, as is often the case, seen by Aboriginal people as a place of significance.

In terms of Aboriginal concepts of Dreaming travels reported in the literature cited here, two giant snakes had passed by here on their way from Noonkanbah, and there were several other dreaming tracks of significance associated with the area. Umpampurru itself was a fertile place from which goannas could be dreamed: goannas were, according to some Aborigines, attracted to the hill from the surrounding area, and a ritual adept could 'dream' them out of the hill and ensure a continuing supply of them to people. The Western Australian state government under Charles Court was anxious to have the company proceed, and in the end the company took the position that it had no choice but to engage in exploratory drilling.

This resulted in the finding that the site was dry, and the drill-hole was then plugged. Objections to the drilling were widely taken up by supporters of the Aboriginal people, who were seen by many as traditionally-oriented and victimised in this episode. But there was also fierce objection to their opposition to the drilling on grounds of 'traditional culture', on the part of Charles Court among others, who said in an angry letter to then Minister for Aboriginal Affairs Senator Fred Chaney that it was impossible to have 'a sensible communication with you to achieve a practical result in respect of Aborigines, their land, and mining and petroleum operations' (quoted in Hawke and Gallagher 1989:314).

In the Noonkanbah case, ways in which problems of Aboriginal relationship to country were to become conventionally debated emerged with some clarity. Among these were the problems created for developers by two related Aboriginal perspectives: first, that all the country is meaningful and important, and areas of significance are not reducible to narrowly defined 'sites'; and second, that places are interconnected, so that a place may be seen as being interfered with at some considerable distance from a development activity. Issues were also raised about what kind of information should be supplied to developers: whether the locations of named places should be mapped and thereby revealed, or whether exclusion zones should simply be indicated, inside of which exploration or other activities were not to occur. A Melbourne *Age* opinion poll of 1980, which asked whether the federal government should have tried to use its powers to stop oil drilling at Noonkanbah, received a 45 per cent positive response (including a high proportion of Liberal voters), as compared with 47 per cent against. At the same time, in Western Australia, amendments were introduced to the *Aboriginal Heritage Act,* which effectively narrowed the definition of an Aboriginal site, and gave rights of appeal under the act to any person with an interest in land, without any similar right of appeal being given to Aboriginal people.

Jabiru

Another high-intensity dispute emerged in the late 1970s around the proposed Ranger Uranium mine site near what has become the town of Jabiru (about 130 kilometres north of Coronation Hill; see Map 1, p. ix). The Northern Land Council had at this time recently been established. In advance of the development, the federal government commissioned efforts to define Aboriginal relations to land in ostensibly originary, traditional terms (of 'clan territories', and so on), as part of a wider investigation of the development project. To me (I was new in the region, and working with socially linked people) it seemed that such traditionalist terms of relationship to land were quite inadequate for dealing with the effects of the regional history of colonial activity and dislocation. More open acknowledgment of these issues might have altered the terms in which identification went on, but this may have been felt by the Northern Land Council to be too risky and potentially disenfranchising. There was almost no doubt that mining would proceed, despite environmentalist and Aboriginal opposition.

A senior Aboriginal man named Toby Gangale was one of a very few persons who had been securely found by the Fox report of 1977, federally commissioned prior to mining development, to be a traditional owner for a delimitable clan territory, encompassing Jidbijidbi, or Mount Brockman, just north of the proposed Ranger mine site. Contention arose as to whether and, more realistically, where, development could proceed, given Aboriginal views of the dangerous character of the locality of Mount Brockman. The previous processes of identification of landowners in quite narrow terms made it possible to quickly pick out Toby Gangale, as senior owner of Mount Brockman, as the man to whom pressures would be applied as part of the definition of this issue in the national media, and he accordingly came under intense national scrutiny. He was pressed for many months to take a position on the acceptability, or otherwise, of mining in relation to Mount Brockman by representatives of environmental, anti-uranium, and pro-mining forces. Some environmentalist groups seemed to hope that a definitive Aboriginal statement could lead to, or at least support, abandonment of the project. Over many months, Toby Gangale generally maintained that mining was undesirable in the immediate area of Mount Brockman, but he had no more definitive idea than anyone else how to resolve the impasse, which he was forced to live out through daily interrogation, between the nature of the place as he represented it, and the highly transformative nature of the proposed project. In the end, the mine was developed at the proposed site with some recognition ostensibly given to Aboriginal concerns about the nature of the place. (His eldest daughter Yvonne is now frequently in the press, having been deeply involved for several years in an effort to prevent uranium mining at near-by Jabiluka).

Coronation Hill

A later dispute developed from 1985 around the issue of whether an earlier-known and partially exploited locality known as Coronation Hill might be the site of a major uranium and heavy metals development (see Map 1, p. ix). Lying within a relatively remote area slated for eventual inclusion within Kakadu National Park, Coronation Hill was originally outside the boundaries as first drawn up of 'Stage Three', so called, of the Park, and within an area oxymoronically labelled the 'Conservation Zone', where mineral exploration and development were to be permitted if finds were prospective. Later, however, boundaries were redrawn so that Coronation Hill was incorporated into the Park, after the reporting of the federally-created Resource Assessment Commission, and prime minister Robert Hawke's recommendation that mining not proceed. Figuring largely throughout the dispute, in addition to objections to mining raised on environmental and other grounds, were Aboriginal claims concerning the dangerous and sacred significance of the area arising from its inhabitation by a male hunter-creator, Bulardemo, capable of apocalyptic destruction if disturbed (Keen and Merlan 1990, Merlan 1991). This case, in which I was closely involved, had a number of complications, though their character was not too surprising.

The general area had been and continued to be one of enormous renown to Aboriginal people of the region because of the Bula story. But knowledge and experience of this fairly remote area was generationally very stratified, and linkage to the area in familial and clan terms was quite restricted. To some extent and for some people, the Bula story had become dislocated, and even those closely linked to the area in familial and clan terms had not spent much time there for decades. There were consequent uncertainties about the identification of places which became significant in a context where the developer wished to reduce the discussion about the area affected to its narrowest possible terms, and in any case, to the rationalised terms of mapping and contract.

The attitudes and dispositions of the three senior Aboriginal men who became identified as the main custodians seemed to me, who knew them all well, complex. They were not prepared to give up the identity and social standing that they had enjoyed life-long as a result of their attachment to this region of apocalyptic renown. Their attachment to the area through family connection and early experience, and its imaginative impress upon them, seemed to me profound. But on the other hand, their lives had been complex. They had always had to live as Aborigines with whitefellas in mind, and their adult lives had been shaped by extended relationships with white bosses, and by experiences in cattle and mining camps and later, on the fringes of regional towns. The spaces from which they might make choices were more diverse than the eventual national identification of them as traditional Aboriginal law-men, driven by religious concerns. The terms in which they were being required to respond – whether they were 'for' or 'against' the development – were stark, and events constantly led to further polarisation. The developer (BHP Gold, later the Coronation Hill Joint Venture) had established some contacts with Aborigines, and gotten some younger men to work drilling cores in an early phase, so that as the dispute developed it was fuelled by intra-familial and inter-generational animosity. This was characterised by the increasing growth of opposition between developer, some Aboriginal proponents of the project, and the Northern Territory government on the one hand, and Aboriginal organisations, some federally sponsored, on the other. The national depiction which gathered strength was of Aboriginal relationship to Coronation Hill as part of an inclusive system of religious values and perspectives on the world.

Hindmarsh Island

Most recently, the proposal to develop a bridge at Hindmarsh Island (see Map 1, p. ix) came to focus on claims concerning the secret and gender-restricted significance of the area at the mouth of the Murray River, as represented by some Ngarrindjeri women who became known as the 'proponents', and denied by other Ngarrindjeri women who came to be known as the 'dissidents' (Kenny 1996). The proponent claim was that building the bridge would fatally impair the reproductivity of women by imposing something between water and sky at this place. As he was empowered to do, the Aboriginal Affairs Minister, Robert Tickner, used the power of the *Commonwealth Aboriginal and Torres Strait Islander Heritage Protection Act (1984)* under section 10(4) to place a twenty-five year ban on the construction of the bridge. However, a royal commission instituted into the case in South Australia in 1995 opined that what was now referred to as 'secret women's knowledge' had originated no earlier than 1993, and there were intimations that the proponent women had been influenced by environmentalists and others. When a new federal minister tabled a special bill in Parliament in 1997 to allow the bridge to be built, the original claimants lodged a High Court appeal, claiming the special bill was a racist refusal of the asserted beliefs of Aboriginal persons. This was dismissed by a High Court decision 5-1 in February 1998, and the *Hindmarsh Island Bridge Act* was found to be valid, and to the extent of inconsistency, to override the Commonwealth act (see Weiner n.d.).

In terms of my general argument, it might be said that the proponent women have repeatedly asserted that they have certain beliefs, and that these are an integral though highly restricted aspect of Ngarrindjeri culture and heritage. But various tribunals have found this claim unproven. Given the denials of the dissident women that they ever heard any of this, doubt has been cast upon the transmission of such beliefs through time, the generality of their distribution, and their existence as part of a demonstrable framework of conceptualisation and social action, or anything one might call contemporary Ngarrindjeri culture (see Weiner n.d). The fact that the proponent argument has been couched in an ostensibly traditional language of myth and landform suggests that the proponent women saw this as the strongest way of putting their case. One of the ways of understanding the aggravated dispute in Hindmarsh is around the question of whether this kind of representation was deliberate strategy and 'invention' in the sense of conscious contrivance, or whether it is to be seen as an expression of Ngarrindjeri culture, albeit under pressure from transformative influences. Clearly, the royal commission finding suggested contrivance. Circumstantially, the proponent women's use of myth and their accounts about landforms demonstrate the hegemony of traditionalist stereotypes, but may leave unresolved the question about the deliberate and strategic nature of their position. It is notable that the women have made constant use of the notion of 'belief', and they, their legal counsel and their anthropologist objected to questions probing their beliefs as oppressive. They argued that they are perhaps not susceptible to the scrutiny of others, on the one hand because they involve secrecy, but on the other, simply because they are 'beliefs' and beyond the right of others to question.

What is being questioned, though, is not the beliefs themselves (which many would take to be false, prima facie), but their epistemological status, that is, whether or not they form part of an identifiable cultural complex, unselfconsciously held rather than strategically adopted. I would observe that these women are clearly operating within a different framework than most Aboriginal people I know in the Northern Territory, who tend to point out to people like myself actual landforms as the physical counterpart of mythic stories, 'so you can believe', as they say. In other words, they are not making use of the notion of belief as interior state, nor relying simply on the idea that 'seeing is believing'. They assume rather that stories and physical manifestations are part of the same larger reality. They confidently rely on the material forms as the demonstration that stories of creator heroes are, as they would say, 'true', the physical counterpart of the stories they tell. These forms of relationship to land have not been seen as characteristic of contemporary Ngarrindjeri. Therefore the proponent argument has been seen as a strategic attempt to reappropriate or reclaim that sort of cultural configuration, rather than as an argument made from within a culture to which that kind of way of doing and seeing things is inherent.

Another cause of dissension was dispute among government officials. There were those such as former Minister for Aboriginal Affairs Tickner, who were originally willing to fully recognise allegedly gender-restricted secrets, by putting them in marked envelopes and not reading the contents, even though the decision to ban the bridge, with all its consequences, rested on acceptance of their validity. There were others however who were not willing to make the assumption that the envelopes contained material of such nature that it would provide a forceful cultural argument against building of the bridge. The willingness to conform to the alleged proprieties of 'another' culture was seen as suspect and even ridiculous by those who did not believe that sort of radical otherness was involved, and the Minister was judged as meekly taking his part in a charade.

If Hindmarsh has been seen as a watershed, it is partly because it has been the case which has raised to poignant visibility 'cultural difference' argumentation itself, rather than just its degree of validity in particular cases, and made its problematically dual character clear. With respect to Australian indigenous issues, heritage legislation is to protect 'their' culture to some extent, but in a way also deemed compatible with limiting and defining their claims upon the public domain. Hindmarsh raises questions about the extent to which the 'difference' which such legislative schemes presuppose is a sustainable basis for decision and action when it is perhaps not radical, does not involve a completely different scheme of values and perspectives than is present among the range within the dominant society, and when some of those involved share with the dominant society some basic, even if tacit, assumptions about the nature of difference.

The proponent women's position went, for many, beyond credibility, demanding acceptance of certain forms of Aboriginal belief and action as the basis for the protection of place when there is widespread suspicion even among Aboriginal sympathisers that other bases or differently formulated bases would be more appropriate, a closer approximation of what 'they' are like at this historical juncture (Bell 1998 notwithstanding). The dispute has highlighted the contradictory tendencies or 'moments' within the dominant society, first, to seek to expunge 'difference' when it is experienced as intolerably great and indomesticable (as in frontier days, or in assimilatory moments); but second, to elicit and valorise difference as that which might justify 'differential' or 'special treatment'. The stark difference between these two moments, and their value-laden quality, makes it difficult to find terms for subtler recognition of cultural differentiation, as well as to debate the possible significance of other than 'radical' difference.

Beyond this lie some bleak prospects: on the one hand, the extreme limitation of the liberal social project as in the recent rise of the One Nation Party and the denial of representation in existing institutional arrangements of any form of 'otherness'. On the other lies the continuation of the division between liberal intellectuals engaged in (what Kahn 1995 calls) an 'expressivist' critique of modernism, and other, more managerially oriented elites, the former recognising themselves as a group focused upon culture and meaning and the defence of difference, versus those professionally oriented towards management, technical mastery and rationalisation (Kahn 1995). We can briefly consider how such a division is presently and conflictually played out, by considering the ways in which something like these two divisions, as represented by anthropologists and developers, deal with the problem of bounding Aboriginal involvement in sacred sites issues.

Expressivism versus techno-managerialism

Sacred sites incidents have generally involved Aboriginal people being presented with development proposals impinging upon them from the outside, and that are quite different in scope and in kind from anything most of them would have imagined taking place. Different from many Papua New Guinea situations, their historical relations with settler society have been such that they do not tend to articulate a clear and optimistic aspiration for 'development', and have not had the opportunity to think of themselves as potentially involved in or benefiting from it (though I must say I have had various Aboriginal people show me alleged gold, oil and copper finds, and urge me to take samples to assay secretly).

Partly because of a great deal of continental commonality in ways in which places are understood to be significant, and partly because of the transformative nature of many of the proposals themselves, it has often appeared as if development will supersede and displace important values Aboriginal people may have associated with a particular place or area. Many kinds of development, such as mining, may range over a wide area in the exploration phase, but later come to focus on a particular location. Discussions with Aboriginal people often require the latter to do something they may be quite reluctant to do: to focus their attention upon places as 'sites', that is, as abstract locations separated from the network of connections which constitute their significance from an Aboriginal point of view. Further, Aboriginal notions of 'place' may have a past orientation, in the specific social sense that Aboriginal people may not, given changes in their forms of life, have occupied the area or place in question for some time, and thus may not have renewed and perhaps modified and updated their sense of its current significance through the everyday kinds of social processes that are continuous and ordinary under conditions of immediate and intimate bodily relationship to country. This 'pastness' may not detract from the significance for them of their previous sense of the place, although there may be questions of the distribution of representations of place, and their currency. Perhaps most significantly, the nature of the proposal made to them about the place *may* be quite different from anything they have previously had to decide about any place of importance, and the alteration proposed may be drastic.

Where development concerns a place about which knowledge and representations are not widely distributed, but also due to the unusual or unforeseen nature of many kinds of developments and alterations, Aboriginal people may be called upon to present a picture of the significance of places that may have to be worked on afresh as a current collective representation, to the extent that it *is* current, or collective. There arises intense and politicised demand for them to say 'what a place signifies' for its possible relevance to the development decision, but without reference to the present circumstances of development or other proposed action.

But to give experience of place a present articulation is inevitably to re-shape it to some degree, or to develop some new sense about its significance relative to the present. Strenuous demand is made of Aboriginal people that they keep separate their 'knowledge' about places from any issues of current proposals relating to them. There is in this demand a recognition that to think consciously about the relevance of one's existing horizons of information and understanding to the present might be to change the terms of one's evaluations, or even to change the terms of one's representations.

There is a possibility for conceptual percolation from what was to what is, and this might introduce unacceptable indeterminacies into the development proposal. On the other hand, developers typically consider it in their interests to discuss the project, and its often gargantuan dimensions, without feeling constrained by the sense in which this may conflict with existing indigenous representations. In this way, contemporaneity and futurity of relationship to place is formulated as a property of development, but not as a property of Aboriginal relationship to place, which is characterised by pastness. Sometimes even the validity of pastness is not conceded, as when the developers in the Coronation Hill case claimed that the Bula complex was 'archaic', no longer current at all. Let me illustrate this kind of tension between developer and indigenous representations as they emerged in the Coronation Hill dispute.

The aim of the developer was to gain permission to explore, and then to mine. Through a long process, they accepted (at least for a time) a notion that permissions, if they were to be gained in any demonstrable sense from principal Aboriginal persons, would best be 'incremental' – a step at a time. They were unlikely to gain permission for all they wanted to do at once. (They always felt matters would go more smoothly if they were allowed to deal 'directly' with the Aboriginal people involved, without interference from Land Councils and the like). They tried demonstrating low-level blasting on the hill in the presence of Aboriginal people, and gained varying and inconsistent degrees of acquiescence for this. At various times they felt encouraged by comments from the three custodians that Coronation Hill was 'all right', 'nothing there', 'him just come past through there' (referring to the creator being), 'it is free' – but there were other statements opposing the works on various bases. The custodian who eventually became strongest in his negative pronouncements usually put matters on the basis of tradition, inheritance and responsibility passed on to him from his father: 'My father's law never allowed people to work down there because they are Bula places; all sacred sites. So I am still going on from what my father said' (Peter Jatbula, Barunga evidence, March 1987).

As against all this temporising, the developers clearly wanted comprehensive permissions. To that end, they made a video of an interview between a geologist working for them, and the eldest male custodian in company with a friend and camp mate (who did not claim to come from this country, and thus was simply 'help' or company to the older man). The interview began with introductions and scene-setting by the geologist. It then featured him gradually dismantling a very realistic-looking, papier mache model of Coronation Hill to gain approval from the two Aboriginal men – a kind of magical enactment, if you like, of the approval process, without the let or hindrance he was actually encountering from a combination of sources. He said something like, 'In the first stage the Hill will look like this', and he removed a section from the model hill. 'Oh, yeah', said the two men. 'And then if we find enough, in the next stage the hill will look like this', he said, removing another slab from the mountain. 'Yeah', said the two men. 'Is that all right?' the geologist asked. 'All right', they echoed. 'And then if we find some more, the next stage will look like this', he said, removing another section and leaving a much reduced model, something that looked less like a hill, 'is that OK?' 'Oh, yeah, good one', said the Aboriginal helper enthusiastically, with (I recall) nods from the older man.

The interaction here looks a bit like theatre. An interactional scene has been set and reconfirmed several times over in the course of a few minutes. Aboriginal participants seem to know what is expected of them and respond accordingly. The video ended with the geologist demonstrating the last phase, with the hill almost levelled, the two men having said 'OK', and sign-off by the geologist. This was a rather graphic imposition of an alternative representation of what might be, over the ones normally entertained by these Aboriginal men.

If this was a whitefella invention to show what might be, the Aboriginal custodians never were in a position to directly impose their representations upon others. But it might be said that they tried to place them underneath others', in keeping with their ideas of the interiority of the sacred in the ground, and to establish more firmly what already was. One of their counter-moves during processes of consultation was to apply forms of existing cosmological convention to the proposed mine-site. To explain: one of the kinds of thinking that provided the background to both their statements of permission and of prohibition was the motif of differences between what custodians referred to as 'Bula places', and the concept of wide-ranging travels by Bula hunting over this entire area of the Upper South Alligator River Valley. 'Bula places', in their vocabulary, were where Bula went into or came out of the ground, and these were held to be very 'dear' and dangerous, and it is unlikely that custodians would have agreed to mining at them (or near them, though this got into difficult matters of degree). But in their minds there was a distinction between such interior places, and the wider area of Bula travels, and a certain ambivalence about how to evaluate and resituate that distinction in terms of the development proposal. (For the significance of underground, see Hawke and Gallagher 1989:125, Kolig 1987:127-30.)

Coronation Hill had not, in my experience, been said to be a Bula place; it was rather a place Bula had hunted through, moved through above ground, and near which he may have scraped along the ground when bitten on the knee by a hornet (though the real-world locale of this story element was not firmly established). But in my experience, these men had long amalgamated imagery from secret-sacred practices formerly involved in Bula ritual, and their notions of this as an area where mineral deposits had previously been found, into a formulation of minerals in this region as the innards, or essence (*ngan-mol-ngayu*) of Bula. Whether or not they may have previously said or thought this of the particular Coronation Hill deposits, I do not know, though they certainly did of the South Alligator Valley in general, and this underpinned their sense of the entire area as *japurru* 'dangerous'.

But the fact is that they seamlessly joined some of the conventional Bula imagery to the present situation of the mining development issue, in saying expressly that the minerals at the Hill were the 'blood' of Bula, and/or of his wives – this was the limited extent to which they underlaid, so to speak, their conventional understandings beneath the development proposals. By certain commentators, this making explicit was called an 'elaboration' (Levitus 1990), by others an outright invention in the sense of contrivance (Brunton 1991c). It was such meaning formations as these that gained recognition as definitive in terms of one of the national-level predispositions that prevailed – the understanding of these men's position as one in which 'religion' was of decisive importance.

How does this kind of dispute look from the perspective of the divisional structure I referred to above, between the intellectual 'expressivist' critique of modernism, versus rationalising techno-managerialism? Let us consider each side in turn. In general, anthropologists as representatives of the intellectual expressivist position, have responded to such conjunctures by reference to some notion of the ongoing and creative character of culture, often referred to in recent literature as 'invention', but without any necessary suggestion of deliberate contrivance. The term seems to have become prominent recently from Hobsbawm and Ranger's (1983) book, *The Invention of Tradition*. The force of this apparently oxymoronic title was to unsettle understandings of the fixity of ethnicities, nations and national traditions. 'Traditions', including nationalist ones, are often thought of as the stable underlying cause of collective action rather than as its 'changeable product' (Calhoun 1991). Hobsbawm and Ranger set out to surprise everybody by demonstrating the recency and mutability of various, apparently well-established 'traditions', such as those surrounding the British monarchy.

Various well-known anthropologists, including Sahlins (1985,

1993, also Kirch and Sahlins 1992), and Wagner (1972, 1975) have made extended historical and ethnographic arguments about the ways in which culture is a continuous product of a dialectic between structure and action, or in Wagner's terms, between convention and invention. Even more recently, further arguments have been made about this problem of the continuous reformation of 'culture' by Jackson (1989), and Briggs (1996), among others. 'Culture' or 'tradition', they all argue, is apparently finished only in historical and synoptic accounts; in actuality, it is continually under construction. But the problem is, how? and when? and is this issue critical in respect to the kinds of contentious development disputes I have been discussing?

A general problem in this literature is the extent of its dedication to the cultural categories of particular societies as the source of creativity, without further consideration of the terms and moment of interaction. To relate that to our previous example: we see that pre-existing Aboriginal ideas about Bula's travels was a presupposition that had lent the wider area of the upper South Alligator Valley an exceptional character, in terms known to many Aboriginal people. Given this, the concept of Bula's travels was also a major source of creative potential around the questions that arose in this dispute, What is the significance of a particular locale in this area? And, What is the significance of precious metals? Where precious metals were affirmed to exist, it might be expected that they would be linked, by Aboriginal people who think in these terms, to the Bula travels; and indeed, the sense of significance linked to Bula had extended (to my knowledge) over the upper South Alligator, and was not sharply delimitable. We still have to recognise that the issue, and the specific claims about Coronation Hill, arose at a particular historical moment; it is insufficient to invoke the notion of 'invention' as continuous cultural construction. One must also recognise the problem of identifying creative potentials in action, and in interaction, in a way that does not simply privilege particular cultural categories, but also places some emphasis on the structure, practice and timing of interaction itself. To this extent, the objecting developers have a case to be considered: Why now, at this juncture, is this identification made? And why was the association with this particular place not previously (and independently) recorded? However, they have often asked those questions from a position of power, and a politics of suspicion (one that may fail to accord recognition and respect to Aboriginal cultural categories and phenomena in any event).

On the other side of the argument is the evident politics arising from this lack of recognition: if Aboriginal people are not seen as having entitlements, if the nature of significances they identify are considered suspect; if their bargaining position is weak and under attack, they are at a disadvantage. They may, wittingly or unwittingly, redress that where possible by asserting the strength and absoluteness of their own cultural resources. Anthropologists are perhaps prone to take the view that their own cultural resources are all that are involved, but also recognise, or complain, that power-laden interaction is involved. The absolutism of whole-culture arguments may emerge as a product of the tense interaction, rather than its being clearly or simply a property of Aboriginal relation to country. Contrasting with the absolutism of Aboriginal positions as these sometimes are stated publicly, is a lability of positions as these emerge in many interactions in which Aboriginal people, privately confronted, appear to accept points of view put to them by interested others.

Let us look at this from the perspective of developers. In these disputes, the basic position of developers with respect to the issues we have been discussing is not very nuanced. Generally, they adopt a consequentialist market morality, asserting that the outcomes of their development activity will fall into patterns of benefit for the 'whole community', so that narrow interests should not be allowed to prevail against it. They have shown variable willingness to take the views of Aborigines and others who may claim disruption or disadvantage into account, to the extent this will not cause them to stray too far from their own imperatives, often quite restricted time frames, and may help to minimise capital exposure risks. For developers, the best possible course is one of limiting the negotiating process to rationalised and/or legal terms, rather than attempting to deal with broad historical or diverse and possibly difficult social justice claims. Large-scale developers can be assured that the state takes an interest in them, in so far as they are crucial to its economic performance. State and developers have a shared interest in their ability to deliver economic benefits and infrastructural services. Thus the state is armed with 'national interest clauses' capable of overturning sectional objections, even as 'nation' and 'state' have become increasingly 'de-hyphenated' under conditions of the hypertrophy of the global economy (Turner 1998). It arguably becomes more difficult to think in terms of unitary 'nations' as the ideological dimension of given 'states'. From the developer's perspective questions are: Is there significance here, in just *this* patch of ground with which I am concerned? and, Since when has this been the case? For them, freedom to operate presupposes the clear definition and bounding of other interests.

The effort at precise temporal and spatial bounding makes me think of parallels elsewhere in our rationalised institutions, for example, in the notion of 'pre-existing condition'. When one wants to take out certain kinds of insurance, to legitimately qualify one must submit to examination to demonstrate that there is no 'pre-existing condition' that would tip the probabilities of having to pay out against the insurer. If one already knows one has a tumour, one is supposed to say so. The condition of possibility that would-be insured know something they are not saying, is referred to by insurers as 'moral hazard'. It is a moral hazard to them for me to suspect or know I have a tumour, and not to say so.

From the perspective of developers, the unrationalised indigenous practices I have exemplified above, and the seemingly unpredictable and unaccountable nature of interactional outcomes, constitute a vast area of 'moral hazard' which may impede their operations. For Aborigines and some others, the job of opposition to development projects has involved attempting to muster sufficient support and to project a position sufficient to overcome the impression of the narrowness or insufficiency of objection. In some of the cases I have cited, there has been some common cause between Aboriginal and environmentalist organisations, but by and large this connection has not been strong. In recent times, Aboriginal groups had found their firmest allies in sections of federal government, and in the politics of national and international embarrassment. But the preference of developers, aside from no interference at all, is for the treatment of all such issues to be subject to forms of rationalisation; for issues that might be socially and culturally defined to become subject to some predictable and scheduled regime. It is exactly this, I would argue, that we are seeing in the slow process of being realised in some recent indigenous affairs. The political liberal strategy involved, I suggest, has been one of revising 'special status' in terms of comparability with dominant institutions. Let me illustrate with Native Title (though other examples could be given).

Rationalising moral hazard

If the Gove case found Aboriginal title to be incomparable and unrecognisible at law, the 1976 *Land Rights Act* was formulated to treat it as incomparable but recognisible under special beneficial statute designed for the purpose. Native Title (originally found in the Mabo decision, 1992) was in some respects the tail-end of the Whitlamite comet of recognition of cultural difference, but in other ways represents a new inflection on the old vision of a culturally differentiated world. It treats native title as potentially having survived, based in some originary status, and requires that for it to be deemed to have survived, continuing systems of 'law and custom' of some kind must be found in terms of which relationship to land has been sustained. But it also finds Native Title to be justiciable at the common law, and in that respect admits it to Anglo-Australian law as never before.

On the one hand, this creates a space for contemporaneity and recognition. I think it is important to take the view, as Noel Pearson (1997) does, that this arrangement, the recognition of a form of law and custom by another as comparable to itself and within its purview, does not essentially treat Native Title as an independent reality, but creates it as a generic 'recognition concept', so that wherever Native Title is found in particular cases, it is an avowedly intercultural product which becomes visible as survival on the one hand, recognition on the other.

But comparability and recognition also create the space for title to be subject to the same kind of rationalised processes as other title. By this token, developers are more inclined to take the view that they have a rule-ordered scheme under which to negotiate with Aboriginal interests, and that avenues for them to make contact with Aboriginal communities and people exist in ways they are more comfortable with. From what I have read and seen, it is a jungle of emergent interests, and the likelihood of increasingly legalistic process in the Native Title and regional agreements era is great. As far as Aboriginal people are concerned, my supposition is that new possibilities for inclusion in negotiations and consultations on some terms, even if from relatively weak bargaining positions and in respect to projects outside their control to veto or affect substantially, will create conditions under which some of their concerns which were previously unpalatable and unpublicisible if they were to be seen as sufficiently culturally different, will now overtly become the basis for positioning and bargaining.

Conclusions

I began by comparing the conditions of current project assessment and mining development in Papua New Guinea and in Australia. I noted that the common Papua New Guinean enthusiasm for development often leads to rapid acceptance of projects at the community level, but that this has, in certain well-known instances, waned or disappeared as the practical entailments and unforeseen consequences of projects manifest themselves. Against this briefly sketched Papua New Guinean background, I have argued that Australian conditions of long-term dispossession, subordination, and assisted recuperation, together with Aboriginal orientations towards country as significance-laden, have meant that there has been comparatively less enthusiasm among Aborigines for 'development' as a good in its own right. With the growth of organisations which have been able to represent Aborigines, there has been a tendency for potential 'goods' to be thought of as culture-bounded, those that are 'Aboriginal' having to do with the preservation of sites, and of (other aspects of) Aboriginal culture, and others which are material, social and developmental to be thought of as flowing from projects which are essentially 'non-Aboriginal'. This is quite a different conception of social goods from the Papua New Guinean cases, where 'development' with a significant material and infrastructural component, emerges strongly and often in advance of any on-the-ground development, as a desire of the indigenous community. I have shown that a good deal of recent dispute in Australia about the development of areas and sites of significance hinged on arguments about their relevance to an 'Aboriginal culture' conceived holistically. Such holistic conceptions of Papuan New Guinean cultures tend to have been secondary developments in reaction to the perceived negative effects of development already undertaken.

The *Native Title Act*, in recognising the possibility that Aborigines hold title in land compatible with the Common Law, makes them people with potentially recognisable relations of right and of property, and thus may render them more accessible to a more familiar politics of negotiation, compensation and price, regulated through legal and governmental institutions. It may seem paradoxical that the enactment appears both to recognise property-holding as something belonging to Aboriginal culture, and makes it subject to closer scrutiny and regulation. This appears less paradoxical if one sees these developments as part of a longer history of accommodation in a settler colony in which acceptance and protection of Aboriginal institutions has always, more or less overtly, also been delimiting and incorporative.

REFERENCES

Aboriginal Sacred Sites Protection Authority. 1982. *Report on the Castlereagh Bay and Howard Island sea closure application*. Vol. I. Darwin: Aboriginal Sacred Sites Protection Authority.

Allen, B. 1995. 'At your peril: studying Huli residence' in *Papuan borderlands: Huli, Duna, and Ipili Perspectives on the Papua New Guinea Highlands*. Edited by A. Biersack. Ann Arbor: University of Michigan Press.

Alpher, B. 1991. *Yir-Yoront lexicon: Sketch and dictionary of an Australian language, trends in linguistics documentation 6*. Berlin, New York: Mouton de Gruyter.

Altman, J. 1983. *Aborigines and mining royalties in the Northern Territory*. Canberra: Australian Institute of Aboriginal Studies.

Attwood, B. 1996. *In the age of Mabo: History, Aborigines and Australia*. Sydney: Allen and Unwin.

Austin-Broos, Diane J. 1996. What's in a time, or a place? Reflections on Swain's hypothesis. Review symposium on Tony Swain's: A Place for Strangers. *Social Analysis* 40:3-11.

Bakhtin, M. 1981. 'Forms of time and of the chronotype in the novel. Notes toward a historical poetics' in *The dialogic imagination. Four essays by M.M. Bakhtin*. Edited by M. Holquist. Austin: University of Texas Press. pp. 84-258.

Ballard, C. 1994. The centre cannot hold: Trade networks and sacred geography in the Papua New Guinea Highlands. *Archeology in Oceania*. 29:130-148.

———. 1995. The death of a great land: Ritual history and subsistence revolution in the Southern Highlands of Papua New Guinea. PhD thesis, Australian National University, Canberra.

———. 1997. '"It's the land, stupid!" The moral economy of resource ownership in Papua New Guinea' in *The governance of common property in the Pacific region*. Edited by P. Larmour. Pacific Policy Paper No. 19.

Canberra: National Centre for Development Studies, Australian National University. pp. 47-66.

Banks, G. and C. Ballard (eds). (1997). *The Ok Tedi settlement: Issues, outcomes and implications.* Canberra: Research School of Pacific and Asian Studies, Australian National University.

Barnes, J. 1967. 'Genealogies' in *The craft of social anthropology.* Edited by A.L. Epstein. Social Science Paperbacks 22. London: Tavistock Publications.

Barth, F. 1971. Tribes and intertribal relations in the Fly headwaters. *Oceania* 41:171-91.

———. 1975. *Ritual and knowledge among the Baktaman of New Guinea.* Oslo: Universitetsforlaget.

———. 1987. *Cosmologies in the making: A generative approach to cultural variation in inner New Guinea.* Cambridge: Cambridge University Press.

Bateson, Gregory 1958 (1936). *Naven.* Stanford: Stanford University Press.

Beck, Ulrich 1987. The anthropological shock: Chernobyl and the contours of the risk society. Translated by John Torpey. *Berkeley Journal of Sociology* 32:153-65.

———. 1992. *Risk society: Towards a new modernity.* Translated by Mark Ritter. Newbury Park: Sage.

Bell, D. 1998. *Ngarrindjeri Wurrawarrin.* Melbourne: Spinifex.

Bercovitch, E. 1989. Disclosure and concealment: A study of secrecy among the Nalumin people of Papua New Guinea. PhD thesis, Dept. of Anthropology, Stanford University.

Berndt, R.M. 1948. Badu, islands of the spirits. *Oceania.* 19(2):94-103.

———. 1949. Secular figures of northeastern Arnhem Land. *American Anthropologist.* 51(2):213-22.

Berndt, R. (ed.). 1982. *Aboriginal sites, rights and resource development.* Perth: University of Western Australian Press.

Berndt, R.M., and C.H. Berndt. 1947. Discovery of pottery in northeastern Arnhem Land. *Journal of the Royal Anthropological Institute.* 77: 133-40.

———. 1954. *Arnhem Land. Its history and its people.* Melbourne: F.W. Cheshire.

———. 1958. 'Aborigines: myths and legends' in *The Australian encyclopaedia.* Vol. I, 2nd ed. Sydney: Grolier Society. pp. 53-5.

———. 1964. *The world of the first Australians: An introduction to the traditional life of the Australian Aborigines.* Sydney: Ure Smith.

Biersack, A. 1992. Short-fuse mining politics in the jet age: From stone to gold at Mt Kare and Porgera. AAA paper, San Francisco, 5 December.

———. 1995a. 'Introduction: The Huli, Duna, and Ipili peoples yesterday and today' in *Papuan Borderlands: Huli, Duna, and Ipili perspectives on the Papua New Guinea Highlands.* Edited by A. Biersack. Ann Arbor: University of Michigan Press.

———. 1995b. 'Heterosexual meanings: Society, economy, and gender among Ipilis' in *Papuan Borderlands: Huli, Duna, and Ipili perspectives on the Papua New Guinea Highlands.* Edited by A. Biersack. Ann Arbor: University of Michigan Press.

———. 1999. The Mount Kare python and his gold: Totemism and ecology in the Papua New Guinea Highlands. *American Anthropologist* 101(1): 68-87.

Biersack, A. (ed.). 1995. *Papuan borderlands.* Ann Arbor: University of Michigan Press.

Blackburn, J. 1971. *Milirrpum v. Nabalco Pty. Ltd. and the Commonwealth of Australia (Gove Land Rights Case).* Sydney: The Law Book Company.

Bloch, M. 1977. The past and the present in the present. *Man.* 12:278-92.

———. 1995. 'People into places: Zafimaniry concepts of clarity' in *The anthropology of landscape: Perspectives on place and space.* Edited by Eric Hirsch and Michael O'Hanlon. Oxford: Oxford University Press. pp. 63-77.

Blowes, R., and D. Trigger. 1998. 'North west Queensland case study: The Century Mine Agreement' in *Regional agreements: Key issues in Australia.* Vol. I, Summaries. Edited by M. Edmunds. Canberra: Australian Institute of Aboriginal & Torres Strait Islander Studies. pp. 23-29.

Bos, Robert 1988. Jesus and the dreaming: Religion and social change in Arnhem Land. PhD thesis, University of Queensland.

Bourdieu, Pierre 1977. *Outline of a theory of practice.* Cambridge: Cambridge University Press.

Brandenstein, C.G. von. 1970. *Narratives from the north-west of Western Australia in the Ngarluma and Jindjiparndi languages.* Canberra: Australian Institute of Aboriginal Studies.

Breisach, Ernst 1983. *Historiography: Ancient, medieval and modern.* Chicago: University of Chicago Press.

Briggs, C. 1996. The politics of discursive authority in research on the 'invention of tradition'. *Cultural Anthropology* 11(4):435-69.

Brown, P., and A. Ploeg (eds). 1997. Change and conflict in Papua New Guinea land and resource rights. *Anthropological Forum* 7(4). Special issue.

Brumbaugh, R. 1990. 'Afek sang: The "old woman" myth of the Mountain Ok' in *Children of Afek: Tradition and Change among the Mountain Ok of Central New Guinea.* Edited by B. Craig and D. Hyndman. Sydney: University of Sydney Press.

Brunton, R. 1991a. Aborigines and environmental myths: Apocalypse in Kakadu. *Environmental Backgrounder* No. 4. Canberra: Institute of Public Affairs.

———. 1991b. Controversy in the sickness country. *Quadrant* (September):16-20.

———. 1991c. *The significance of shallow tradition: The resource assessment commission on Aboriginal interests in Kakadu.* Canberra: Institute of Public Affairs.

———. 1992. Mining credibility: Coronation Hill and the anthropologists. *Anthropology Today* 8:2-5.

———. 1996. The Hindmarsh Island Bridge and the credibility of Australian anthropology. *Anthropology Today* 12(4):2-7.

Burton, J. 1991. 'Social mapping' in *Customary land tenure: Registration and decentralisation in Papua New Guinea.* Edited by P. Larmour. Monograph No. 29. Port Moresby: National Research Institute.

Busse, M. 1991. Report of research among Fasu speakers, 29 March 1991 to 29 April 1991. Papua New Guinea National Museum.

Calhoun, C. 1991. 'The problem of identity in collective action' in *Macro-micro linkages in sociology.* Edited by J. Huber. Newbury Park: Sage. pp. 51-75

Cape York Peninsula Land Use Strategy. 1993. *What is a CYPLUS?* Queensland: CYPLUS.

Casey, Edward 1996. 'How to get from space to place in a fairly short stretch of time: Phenomenological prolegomena' in *Senses of place.* Edited by Steven Feld and Keith Basso. Santa Fe: School of American Research. pp. 13-52.

———. 1997. *The fate of place: A philosophical history.* Berkeley: University of California Press.

Cawte, J. 1993. *The universe of the Warramirri. Art, medicine and religion in Arnhem Land.* Sydney: University of New South Wales.

Clark, J. 1993. Gold, sex, and pollution: Male illness and myth at Mt Kare, Papua New Guinea. *American Ethnologist* 20(4):742-57.

———. 1994. Shit beautiful: Tambu and Kina revisited. *Oceania* 65(3): 195-211.

Coombs, H. 1978. *Kulinma.* Canberra: Australian National University Press.

Connell, J., and R. Howitt (eds). 1991. *Mining and indigenous peoples in Australasia.* Sydney: Sydney University Press.

Cook, A.C. 1996a. 'Appendix C: mines and mining projects in Papua New Guinea' in *The economy of Papua New Guinea, 1996 report.* Canberra:

Australian Agency for International Development. pp. 153-72.

———. 1996b. 'Appendix D: oil and gas in Papua New Guinea' in *The economy of Papua New Guinea, 1996 report*. Canberra: Australian Agency for International Development. pp. 173-202.

Cordell, J. 1991. Negotiating sea rights. *Cultural Survival Quarterly* 15(2): 5-9.

Cousins, David, and John Nieuwenhuysen. 1984. *Aborigines and the mining industry.* Sydney: George Allen and Unwin.

CRA Limited. 1992. *The history of the Mt Kare dispute.*

Craig, B., and D. Hyndman (eds). 1990. *The children of Afek*. Sydney: Oceania Monographs.

Craig, Ruth. 1969. 'Marriage among the Telefolmin' in *Pigs, pearlshells , and women*. Edited by R.M. Glasse and M.J. Meggitt. Englewood Cliffs, N.J.: Prentice Hall.

Dana, E.S. 1949. *A textbook of mineralogy, with an extended treatise on crystallography and physical mineralogy.* 4th ed. New York: John Wiley & Sons.

Deleuze, G., and F. Guattari. 1983. *Anti-Oedipus*. London: Athlone Press.

———. 1987. *A thousand plateaus*. Minneapolis: University of Minnesota Press.

Department of Aboriginal Sites 1994. *Guidelines for Aboriginal heritage assessment in Western Australia*. Perth: Aboriginal Affairs Department.

Dondorp, A., and B. Uringa 1999. Onobasulu social organization paper. Unpublished MS. Ukarumpa: Summer Institute of Linguistics

Donigi, P. 1988. *The state and property rights in Papua New Guinea*. Port Moresby.

Durkheim, Emile. 1915. *The elementary forms of the religious life*. London: George Allen & Unwin.

Earl, G.W. 1842. Notes on northern Australia and the neighbouring seas. *Journal of the Royal Geographical Society of London*. 12:139-41.

Edmunds, M. 1989. *They get heaps: A study of attitudes in Roebourne Western Australia*. Canberra: Australian Institute of Aboriginal Studies.

Edmunds, M. (ed.). 1998. *Regional agreements: Key issues in Australia*. Vol. 1, Summaries; vol. 2, papers. Canberra: Australian Institute of Aboriginal Studies.

Eliade, M. 1958. *Birth and rebirth: The religious meanings of initiation in human culture*. New York: Harper.

Elkin, A.P. 1953. Arnhem Land music. *Oceania* 24(2):81-109.

Ernst, T. n.d. Social mapping and incorporated land groups project report. Bathurst: Charles Stuart University.

———. 1993. Kutubu Petroleum Project: Social mapping and incorporated

land groups project report. Unisearch (PNG) Pty. Ltd.
———. 1996. On the misplaced concreteness of genealogy (with notes on the clan as fetish). Paper presented to the Anthropology Seminar, Australian National University.
———. 1999. Land, stories, and resources: Discourse and entification in Onabasulu modernity. *American Anthropologist* 101:88-97.
Evans-Pritchard, E.E. 1929. The morphology and function of magic. *American Anthropologist* 31:619-41.
Feld, Steven. 1982. *Sound and sentiment.* Philadelphia: University of Pennsylvania Press.
———. 1996. 'Waterfalls of song: An acoustemology of place resounding in Bosavi, Papua New Guinea' in *Senses of place.* Edited by Steven Feld and Keith Basso, 91-135. Santa Fe: School of American Research Press.
Feld, Steven, and Keith Basso (eds). 1996. *Senses of place.* Santa Fe: School of American Research Press.
Filer, C. 1990. 'The Bougainville rebellion, the mining industry, and the process of social disintegration in Papua New Guinea' in *The Bougainville Crisis.* Edited by R.J. May and M. Spriggs. Bathurst: Crawford House Press.
———. 1992. The management of ethnographic work around the mining industry in Papua New Guinea. Paper delivered at the meetings of the American Anthropological Association, San Francisco.
———. 1993. The policy and methodology of social impact mitigation in the mining industry. Paper presented to the 20th Waigani Seminar.
———. 1996. 'The social context of renewable resource depletion in Papua New Guinea' in *Resources, Nations and Indigenous Peoples: Case Studies from Australasia, Melanesia and Southeast Asia.* Edited by R. Howitt et al. Oxford: Oxford University Press.
———. 1997a. 'Compensation, rent and power in Papua New Guinea' in *Compensation for resource development.* Edited by Susan Toft. Canberra: Australian National University.
———. 1997b. 'Resource rents: Distribution and sustainability' in *Papua New Guinea: A 20/20 vision.* Edited by Ila Temu. Pacific Policy Paper 20. Canberra: National Centre for Development Studies, Australian National University. pp. 222-60.
Fingleton, J. n.d. Laws and procedures for land group incorporation. Report prepared for Chevron Kutubu Oil Project. Canberra: Palata Pty Ltd.
Flinders, M. 1814. *A voyage to Terra Australis.* London: G. and W. Nichol.
Foster, Robert 1995. *Social reproduction and history in Melanesia: Mortuary*

ritual, gift exchange, and custom in the Tanga Islands. Cambridge: Cambridge University Press.

Frankel, S. 1986. *The Huli Response to Illness.* Cambridge: Cambridge University Press.

Frykman, Jonas, and Orvar Löfgren. 1987. *Culture builders: Historical ethnography.* Translated by Alan Crozier. Brunswick: Rutgers University.

Gardner, D. 1981. Cult ritual and social organisation among the Mianmin. PhD thesis, Australian National University, Canberra.

———. 1983. Performativity in ritual: the Mianmin case. *Man* 18: 346-60.

———. 1987. Spirits and conceptions of agency among the Mianmin of Papua New Guinea. *Oceania* 57 (3):161-77.

Gell, Alfred. 1992a. *The anthropology of time: Cultural construction of temporal maps and images.* Oxford: Berg Press.

———. 1992b. 'The technology of enchantment and the enchantment of technology' in *Anthropology, art and aesthetics.* Edited by J. Coote and A. Shelton. Oxford: Oxford University Press.

———. 1995. 'The language of the forest: Landscape and phonological iconism in Umeda' in *The anthropology of landscape: Perspectives on place and space.* Edited by Eric Hirsch and Michael O'Hanlon. Oxford: Oxford University Press. pp. 232-54.

Gerritsen, R., and M. MacIntyre. 1991. 'Dilemma of distribution: The Misima Goldman, Papua New Guinea' in *Mining and indigenous peoples in Australasia.* Edited by J. Connell and R. Howitt. Sydney: Sydney University Press. pp. 35-54.

Gewertz, Deborah, and Frederick Errington. 1991. *Twisted histories, altered contexts: Representing the Chambri in a world system.* New York: Cambridge University Press.

Glasse, R.M. 1968. *Huli of Papua: A cognatic descent system.* Paris: Mouton & Co.

———. 1995. 'Time belong *Mbingi*: religious syncretism and the pacification of the Huli' in *Papuan borderlands.* Edited by A. Biersack. Ann Arbor: University of Michigan Press. pp. 57-86.

Goldman, L. 1983. *Talk never dies: the language of Huli disputes.* London: Tavistock.

Gow, Peter. 1995. 'Land, people, and paper in Western Amazonia' in *The anthropology of landscape: Perspectives on place and space.* Edited by Eric Hirsch and Michael O'Hanlon. Oxford: Oxford University Press. pp. 43-62.

Guddemi, P. 1997. Continuities, contexts, complexities, and transformations:

local land concepts of a Sepik people affected by mining exploration. *Anthropological Forum* 7(4):629-48.

Gupta, Akhil, and James Ferguson. 1997. 'After "peoples and cultures"' in *Culture, power, place: Explorations in critical anthropology.* Edited by Akhil Gupta and James Ferguson. Durham: Duke University Press. pp. 1-29.

Haley, N. 1996. Revisioning the past, remembering the future: Duna accounts of the world's end. *Oceania* 66(4):278-85.

Hammar, L. 1992. Sexual transactions on Daru: With some observations on the ethnographic enterprise. *Research in Melanesia* 16:21-54.

Harris, G.T. 1972. Labour supply and economic development in the Southern Highlands. *Oceania* 43(2):123-39.

Harrison, Simon. 1990. *Stealing people's names: History and politics in a Sepik river cosmology.* Cambridge: Cambridge University Press.

Harrisson, T., and J. O'Connor. 1969. 'Delta iron in a southeast Asian context' in *Excavations of a prehistoric iron industry in West Borneo.* Vol. 2. *Associated artifacts and ideas.* Data Paper No. 72. Ithaca, NY: South East Asia Program, Cornell University.

Harvey, David. 1996. *Justice, nature and the geography of difference.* Malden, MA: Blackwell Press.

Hawke, S., and M. Gallagher. 1989. *Noonkanbah: Whose land, whose law.* Fremantle: Fremantle Arts Centre.

Henningham, S., and R.J. May (eds). 1992. *Resources, development and politics in the Pacific Islands.* Bathurst: Crawford House Press.

Henton, D. 1988. Mt. Kare landownership study.

Hiatt, Lester R. 1975. 'Introduction' in *Australian Aboriginal mythology: Essays in honour of W.E.H. Stanner.* Edited by L.R. Hiatt. Canberra: Australian Institute of Aboriginal Studies. pp. 1-23.

———. 1984. 'Swallowing and regurgitation in Australian myth and ritual' in *Religion in Aboriginal Australia.* Edited by M. Charlesworth, H. Morphy, D. Bell and K. Maddock. St Lucia: University of Queensland Press.

Hirsch, Eric. 1995. 'Introduction. Landscape: Between place and space' in *The anthropology of landscape: Perspectives on place and space.* Edited by Eric Hirsch and Michael O'Hanlon. Oxford: Oxford University Press. pp. 1-30.

———. 1996. Between mission and market: Events and images in a Melanesian society. *Man* (NS) 29:689-711.

Hobsbawm, E., and T. Ranger. 1983. *The invention of tradition.* Cambridge: Cambridge University Press.

Holthouse, H. 1991. *Australian geographic book of Cape York.* NSW: Australian Geographic.

Howitt, R., J. Connell, and P. Hirsch. 1996. *Resources, nations and indigenous peoples: Case studies from Australasia, Melanesia and Southeast Asia.* Oxford: Oxford University Press.

Hughes, Diane Owen. 1995. 'Introduction' in *Time: Histories and ethnologies.* Comparative Studies in Society and History Book Series. Edited by Diane Owen Hughes and Thomas R. Trautmann. Ann Arbor: University of Michigan.

Hyndman, D., and G. Morren. 1990. 'Human ecology of the Mountain-Ok of central New Guinea: a regional and inter-regional approach' in *The children of Afek.* Edited by Barry Craig and David Hyndman. Sydney: Oceania Monographs. pp. 9-26.

Ingold, Tim. 1996. 'Hunting and gathering as ways of perceiving the environment' in *Defining nature: Culture, ecology and domestication.* Edited by R.F. Ellen and K. Fukui. Oxford: Berg. pp. 117-55.

Jackson, J. 1989. Is there a way to talk about making culture without making enemies? *Dialectical Anthropology* 14:127-43.

Jackson, R. 1982. *Ok Tedi: The pot of gold.* Waigani: Word Publishing.

———. 1989. New policies in sharing mining benefits in Papua New Guinea: a note. *Pacific Viewpoint* 30:86-93.

———. 1991. 'Not without influence: Villages, mining companies, and government in Papua New Guinea' in J. Connell and R. Howitt (eds.), *Mining and indigenous peoples in Australia.* Sydney: Sydney University Press.

Jenkins, C. 1994. *National study of sexual and reproductive knowledge and behaviour in Papua New Guinea.* Papua New Guinea Institute of Medical Research Monograph Series.

Jones, R. 1969. Firestick farming. *Australian Natural History* 16:224-8.

Jorgensen, D. 1980. What's in a name: The meaning of meaninglessness in Telefolmin. *Ethos* 8(4):349-66.

———. 1981a. 'Life on the fringe: history and society in Telefolmin' in *The plight of peripheral people in Papua New Guinea.* Edited by R. Gordon. Cambridge, MA: Cultural Survival. pp. 59-79.

———. 1981b. Taro and arrows: Order, entropy, and religion among the Telefolmin. PhD thesis, Anthropology, University of British Columbia.

———. 1983. Mirroring nature? Men's and women's models of conception in Telefolmin. *Mankind* 14(1):57-65. Special issue, *Concepts of conception: Procreation ideologies in Papua New Guinea.* Edited by D. Jorgensen.

———. 1985. 'Femsep's last garden: a Telefol response to mortality' in *Aging and its transformations: Moving toward death in Pacific societies.* Edited by D.E.A. Counts and D. Counts. ASAO Monograph No. 10. Lanham, MD: University Press of America. pp. 203-21.

———. 1988. From sister exchange to 'daughter-as-tradestore': Money and marriage in Telefolmin. *Catalyst* 18:255-79.

———. 1990a. 'The *Telefolip* and the architecture of ethnic identity in the Sepik headwaters' in *Children of Afek: Tradition and change among the Mountain-Ok of central New Guinea*. Edited by B. Craig and D. Hyndman. Oceania Monograph 40. Sydney: Oceania Publications.

———. 1990b. Placing the past and moving the present: Myth and contemporary history in Telefolmin. *Culture* 10(2):47-56.

———. 1991*a*. 'Telefolmin' in *Encyclopedia of world cultures*. Vol. 2, *Oceania*. Edited by Terence E. Hays. Boston: G.K. Hall. pp. 321-4.

———. 1991b. 'Big men, great men and women: Alternative logics of gender difference' in *Big men and great men: Personifications of power in Melanesia*. Edited by Maurice Godelier and Marilyn Strathern. Cambridge: Cambridge University Press.

———. 1993. Magalim goes to work: History, mining and ideas about nature in Telefolmin. Paper presented at the Annual Meetings of the American Anthropological Association, Washington, DC.

———. 1996. 'Regional history and ethnic identity in the hub of New Guinea: The emergence of the Min' in *Regional histories in the western Pacific*. Edited by J. Barker and D. Jorgensen. *Oceania* 66(3):189-210. Special issue.

———. 1997. Who and what is a landowner? Mythology and marking the ground in a Papua New Guinea mining project. *Anthropological Forum* 7(4):599-628.

Kahn, J. 1995. *Culture, multiculture, postculture*. London: Sage.

Keen, Ian. 1978. One ceremony, one song: An economy of religious knowledge among the Yolngu of northeast Arnhem Land. PhD thesis, Australian National University, Canberra.

———. 1992. Undermining credibility: Advocacy and objectivity in the Coronation Hill debate. *Anthropology Today* 8:6-9.

———. 1993. Aboriginal beliefs vs. mining at Coronation Hill: The containing force of traditionalism. *Human Organization* 52(4):344-55.

———. 1994. *Knowledge and secrecy in an Aboriginal religion. Yolngu of northeast Arnhem Land*. Oxford: Clarendon Press.

Keen, I., and F. Merlan. 1990. *The significance of the conservation zone to Aboriginal people*. Canberra: Australian Government Printing Service.

Kenny, C. 1996. *Women's business*. Potts Point, NSW: Duffy and Snellgrove.

Key, C.A. 1969. Archaeological pottery in Arnhem Land. *Archaeology and Physical Anthropology in Oceania*. 4:103-6.

Kienzle, W., and S. Campbell. 1938. Notes on the natives of the Fly and Sepik River headwaters, New Guinea. *Oceania* 8:463-81.

Kirch, P.V., and M. Sahlins. 1992. *Anahulu: The anthropology of history in the kingdom of Hawaii.* Chicago: University of Chicago.

Kirsch, Stuart. 1991. The Yonggom of New Guinea: An ethnography of sorcery, magic and ritual. PhD thesis, University of Pennsylvania.

———. 1995. Social impact of the Ok Tedi mine on the Yonggom villages of the North Fly, 1992. *Research in Melanesia* 19:23-102. (Originally produced in 1993 as Ok-Fly Social Monitoring Program, Report No. 5. Port Moresby: Unisearch PNG.)

———. 1996. Anthropologists and global alliances (comment). *Anthropology Today.* August 12. (4):14-16.

———. 1997a. 'Indigenous response to environmental impact along the Ok Tedi' in *Compensation for resource development.* Edited by Susan Toft. Canberra: Australian National University. pp. 143-55.

———. 1997b. 'Is Ok Tedi a precedent? Implications of the settlement' in *The Ok Tedi settlement: Issues, outcomes and implications.* Edited by Glenn Banks and Chris Ballard. Pacific Policy Paper 25. Canberra: Resource Management in Asia-Pacific and National Centre for Development Studies. pp. 118-40.

Knauft, B. 1995. 'Agency' in cultural anthropology in the mid-90s: A commentary. Paper presented at the Annual Meetings of the American Anthropological Association.

———. 1996. *Genealogies for the present in cultural anthropology.* London: Routledge.

———. 1997. Gender identity, political economy, and modernity in Melanesia and Amazonia. *Journal of the Royal Anthropological Institute* 3:233-59.

Kolig, E. 1987. *The Noonkanbah story: Profile of an Aboriginal community in Western Australia.* Dunedin: University of Otago Press.

Kulick. D. 1992. *Language shift and cultural reproduction.* Cambridge: Cambridge University Press.

Langlas, C., and J.F. Weiner. 1988. 'Big-Men, population growth, and longhouse fission among the Foi, 1965-79' in *Mountain Papuans: Historical and comparative perspectives from New Guinea fringe Highlands societies.* Edited by J.F. Weiner. Ann Arbor: University of Michigan Press.

Larmour, P. (ed.). 1997. *The governance of common property in the Pacific region.* National Centre for Development Studies Pacific Policy Paper No. 19. Canberra: NCDS, Australian National University.

Lattas, A. 1992. Skin, personhood and redemption: The double self in West New Britain cargo cults. *Oceania* 63:27-54.

———. 1993. Sorcery and colonialism: Illness, dreams and death as political languages in West New Britain. *Man* 28(1):51-78.

———. 1996. Colonialism, Aborigines and the politics of time and space: The placing of strangers and the placing of oneself. Review symposium on Tony Swain's *A Place for Strangers. Social Analysis* 40:30-11.

Lawrence, Peter. 1964. *Road belong cargo: A study of the cargo movement in the Southern Madang district New Guinea.* Manchester: Manchester University Press.

Lévi-Strauss, C. 1987. *Anthropology and myth: Lectures 1951-1982.* Translated by Roy Willis. Oxford and New York: Basil Blackwell.

Levitus, R. 1990. 'Historical perspective' in *Aboriginal Areas Protection Authority Submission 77 to the Kakadu Conservation Zone Inquiry.* Canberra: Resource Assessment Commission.

Macknight, C.C. 1972. Macassans and Aborigines. *Oceania.* 42(4):283-321.

———. 1976. *The voyage to Marege. Macassan trepangers in northern Australia.* Melbourne: Melbourne University Press.

———. (ed.). 1979. *The farthest coast. A selection of writings relating to the history of the northern coast of Australia.* Carlton: Melbourne University Press.

———. 1986. Macassans and the Aboriginal past. *Archaeology in Oceania.* 21:69-75.

McConnel, U. 1930. The Wik-Munkan tribe of Cape York Peninsula. *Oceania* 1(1):97-104, 1(2):181-205.

———. 1957. *Myths of the Munkan.* Melbourne: Melbourne University Press.

McDowell, Nancy. 1988. A note on cargo cults and cultural constructions of change. *Pacific Studies* 11(2):121-34.

McIntosh, I.S. 1995. 'Yolngu sea rights in *Manbuynga ga Rulyapa* (Arafura Sea) and the Indonesian connection' in *Native Title: Emerging issues for research, policy and practice.* Edited by J. Finlayson and D.E. Smith. Research Monograph No. 10. Canberra: Centre for Aboriginal Economic Policy Research, Australian National University. pp. 9-22.

———. 1996. Can we be equal in your eyes? A perspective on reconciliation from Yolngu myth and history. PhD thesis, Northern Territory University, Darwin.

Maddock, K. 1970. Imagery and social structure at two Dalabon rock art sites. *Anthropological Forum* 2:444-6.

———. 1991. Metamorphosing the sacred in Australia. *Australian Journal of Anthropology* 2(2):213-32.

Martin, D. 1995. *Money, business and culture: Issues for Aboriginal economic policy.* Discussion Paper No. 101. Canberra: Centre for Aboriginal Economic Policy Research, Australian National University.

Martin, D., and J Finlayson. 1996. *Linking accountability and self-determination in Aboriginal organisations.* Discussion Paper No. 116. Canberra: Centre for Aboriginal Economic Policy Research, Australian National University.

Mathew, J. 1899. *Eaglehawk and crow: A study of the Australian Aborigines, including an inquiry into their origin and a survey of Australian languages.* Melbourne: Melville, Mullen and Slade.

May, R.J., and M. Spriggs. 1990. *The Bougainville crisis.* Bathurst: Crawford House Press.

Merlan, F. 1991. The limits of cultural constructionism: The case of Coronation Hill. *Oceania* 61:1-12.

———. 1997. 'Fighting over country: four commonplaces' in *Fighting over country: Anthropological perspectives.* Edited by D. Smith and J. Finlayson. Canberra: Centre for Aboriginal Economic Policy Research, Australian National University. pp. 1-14.

———. 1998 *Caging the rainbow.* Honolulu: University of Hawai'i Press.

Modjeska, N. 1982. 'Production and inequality: Perspectives from central New Guinea' in *Inequality in New Guinea Highlands societies.* Edited by A. Strathern, Cambridge: Cambridge University Press.

Moore, H. 1994. *A passion for difference: Essays in anthropology and gender.* Cambridge: Polity Press.

Morphy, F. 1983. 'Djapu: A Yolngu dialect' in *Handbook of Australian languages.* Vol. 3. Edited by R.M.W Dixon. Canberra: Australian National University Press. pp. 1-188.

Morphy, H. 1984. *Journey to the crocodile's nest.* Canberra: Australian Institute of Aboriginal Studies.

———. 1990. Myth, totemism and the creation of clans. *Oceania* 60(4): 312-28.

———. 1993 'Colonialism, history and the construction of place: The politics of landscape in Northern Australia' in *Landscape: Politics and perspectives.* Edited by Barbara Bender. Oxford: Berg. pp. 205-43.

———. 1995. 'Landscape and the reproduction of the ancestral past' in *The anthropology of landscape: Perspectives on place and space.* Edited by Eric Hirsch and Michael O'Hanlon. Oxford: Oxford University Press. pp. 184-209.

Morphy, H., and F. Morphy. 1984. The 'myths' of Ngalakan history: Ideology and images of the past in northern Australia. *Man.* 19(3):459-78.

Morren, G. 1979. Seasonality among the Miyanmin: Wild pigs, movement, and dual kinship organization. *Mankind* 12:1-12.

———. 1986a. *The Miyanmin: Human ecology of a Papua New Guinea society.* Ann Arbor: UMI Research Press.

———. 1986b. A small footnote to the "Big Walk". *Oceania* 52:39-65.

———. 1991a. 'Miyanmin' in *Encyclopedia of world cultures.* Vol. 2, *Oceania.* Edited by Terence E. Hays. Boston: G.K. Hall. pp. 209-12.

———. 1991b. 'The ancestresses of the Miyanmin and Telefolmin: Sacred and mundane definitions of the fringe in the Upper Sepik' in *Man and a half: Essays in Pacific anthropology and ethnobiology in honour of Ralph Bulmer.* Edited by Andrew Pawley. Memoir No. 48. Auckland: Polynesian Society. pp. 298-305.

Mountford, C.P. 1956-64. *Records of the American Australian scientific expedition to Arnhem Land.* Vol. 1, *Art, myth, and symbolism.* Melbourne: Melbourne University Press.

Munn, Nancy. 1970. 'The transformation of subjects into objects in Walbiri and Pitjantjatjara myths' in *Australian Aboriginal anthropology.* Edited by R.M. Berndt. Perth: University of Western Australia Press.

———. 1986. *Walbiri iconography.* Chicago: University of Chicago Press.

———. 1996. 'An essay on the symbolic construction of memory in the Kaluli Gisalo' in *Cosmos and society in Oceania.* Edited by Daniel de Coppet and André Iteanu, London: Berg.

Myers, F. 1986. *Pintupi country, Pintupi self: Sentiment, place and politics among Western Desert Aborigines.* Washington, DC: Smithsonian Institution Press.

Nash, J. 1979. *We eat the mines and mines eat us.* New York: Columbia University Press.

Nihill, M. 1994. New women and wild men: 'Development', changing sexual practice and gender in Highland Papua New Guinea. *Canberra Anthropology* 17(2):48-72.

O'Hanlon, M. 1995. Modernity and the graphicalization of meaning: New Guinea Highland shield design in historical perspective. *Journal of the Royal Anthropological Institute* (NS) 1:469-93.

Panoff, F. 1970. Food and faeces: A Melanesian rite. *Man* 5(2):237-52.

Pearson, N. 1997. 'The concept of Native Title at common law' in *Our land is our life: Land rights past, present and future.* Edited by G. Yunupingu. St Lucia: University of Queensland. pp. 150-61.

Peterson, Nicolas. 1973. Camp site location amongst Australian hunter-gatherers: Archaeological and ethnographic evidence for a key determinant. *Archaeology and Physical Anthropology in Oceania* 8(3):173-93.

———. 1986. *Australian Aboriginal territorial organisation.* Oceania Monograph No. 30. Sydney: Oceania Publications.

Peterson, N., and M. Langton. 1983. *Aborigines, land and land rights.* Canberra: Australian Institute of Aboriginal Studies.

Polier, N. 1996. Of mines and Min: modernity and its malcontents in Papua New Guinea. *Ethnology* 35:1-16.

Radcliffe-Brown, A. 1930. The rainbow serpent myth in south-east Australia. *Oceania* 1:342-7.

Rappaport, Roy. 1993. The anthropology of trouble (distinguished lecture in general anthropology). *American Anthropologist* 95(2):247-317.

Reid, A. 1983. The rise of Macassar. *Rima.* 17:117-60.

Reynolds, H. 1990. *The other side of the frontier.* Melbourne: Penguin.

Ridd, Michael. 1997. Impacts of the Ok Tedi Mine and planned mitigation. Unpublished seminar presentation at the conference, The Ok Tedi Settlement: Issues, outcomes and implications, May, Australian National University, Canberra.

Robbins, J. 1995. Dispossessing the spirits: Christian transformations of desire and ecology among the Urapmin of Papua New Guinea. *Ethnology* 34:211-24.

Robinson, R. 1956. *The feathered serpent.* Sydney: Edwards and Shaw.

Róheim, Géza 1925. *Australian totemism: A psychoanalytic study in anthropology.* London: Allen and Unwin.

———. 1945. *The eternal ones of the dream.* New York: International Universities Press.

Rose, D.B. 1984. The saga of Captain Cook: Morality in Aboriginal and European law. *Australian Aboriginal Studies.* 2:24-39.

———. 1988. Exploring an Aboriginal land ethic. *Meanjin* 3:378-87.

———. 1992. *Dingo makes us human: Life and land in an Aboriginal Australian culture,* Cambridge: Cambridge University Press.

———. 2001. 'Sacred site, ancestral clearing, and environmental ethics' in *Emplaced myth: The spatial and narrative dimensions of knowledge in Australia and Papua New Guinea.* Edited by A. Rumsey and J. Weiner. Honolulu: University of Hawai'i Press.

Rudder, John. 1993. Yolngu cosmology. PhD thesis, Australian National University, Canberra.

Rumsey, Alan. 1994. The dreaming, human agency and inscriptive practice. *Oceania* 65:116-30.

Rutz, Henry J. 1992. 'Introduction. The idea of a politics of time' in *The politics of time.* Edited by Henry J. Rutz. American Ethnological Society Monograph Series No. 4. Washington, DC: American Anthropological Association. pp. 1-17.

Ryan, P. 1991. *Black bonanza: A landslide of gold.* South Yarra: Hyland House

Press.

Sahlins, M. 1981. *Historical metaphors and mythical realities: Structure in the early history of the Sandwich Islands Kingdoms.* Association for the Study of Anthropology in Oceania, Special Publication No. 1. Ann Arbor: University of Michigan, Michigan.

———. 1985. *Islands of history.* Chicago: University of Chicago.

———. 1993. Goodbye to *triste tropes:* Ethnography in the context of modern world history. *Journal of Modern History* 65:1-25.

Schieffelin, Edward L. 1976. *The sorrow of the lonely and the burning of the dancers.* New York: St Martins Press.

———. 1991. 'The Great Papuan Plateau' in *Like people you see in a dream: First contact in six Papuan societies.* Edited by E. Schieffelin and R. Crittenden. Stanford: Stanford University Press.

Schieffelin, E., and R. Crittenden (eds). 1991. *Like people you see in a dream: First contact in six Papuan societies.* Stanford: Stanford University Press.

Schutz, A. 1962. *Collected papers 1: The problem of social reality.* Edited by M. Natanson. The Hague: Martinus Nijnof.

———. 1967. *The phenomenology of the social world.* Translated by G. Walsh and F. Lehnert. Evanston: Northwestern University Press.

Sharp, L. 1937. The social anthropology of a totemic system of north Queensland. PhD thesis, Harvard University.

———. 1952. 'Steel axes for stone age Australians,' in *Human problems in technological change: A casebook.* Edited by E.H. Spicer. New York: Russell Sage Foundation.

Sinden, J. 1972. *The natural resources of Australia: Prospects and problems for development.* Sydney: Angus and Robertson.

Smith, D., and J. Finlayson (eds). 1997. *Fighting over country: Anthropological perspectives.* Centre for Aboriginal Economic Policy Research Monograph No. 12. Canberra: CAEPR, Australian National University.

Smith, M. 1994. *Hard times on Kairiru Island.* Honolulu: University of Hawai'i Press.

Sommer, B., and E. Sommer. 1967. 'Kunjen pronouns and kinship' in *Pacific Linguistics Papers in Australian Linguistics.* Series A, No. 10. Canberra: Australian National University. pp 53-9.

Spencer, Sir W. Baldwin, and F.J. Gillen. 1899. *The native tribes of central Australia.* London: MacMillan.

Stanner, W.E.H. 1966. *On Aboriginal religion.*: Sydney: Oceania Monographs

———. 1979. 'Religion, totemism and symbolism' in *White man got no dreaming: Essays 1938-1973.* Edited by W.E.H. Stanner. Canberra: Australian

National University Press. pp. 106-43.

———. 1984. 'Religion, totemism and symbolism' in *Religion in Aboriginal Australia*. Edited by M. Charlesworth, H. Morphy, D. Bell and K. Maddock. St Lucia: University of Queensland Press. pp. 137-72.

Stephens, Sharon. 1995. 'The cultural fallout of Chernobyl radiation in Norwegian Sami regions: Implications for children' in *Children and the politics of culture*. Edited by Sharon Stephens. Princeton: Princeton University Press.

Stewart, R.G. 1992. *Coffee: The political economy of an export industry in Papua New Guinea*. Boulder: Westview.

Strang, V. 1997. *Uncommon ground: Cultural landscapes and environmental values*. Oxford: Berg.

Strathern, A.J. 1982. The division of labor and social change. *American Ethnologist* 9:307-19.

———. 1984. *A line of power.* London: Tavistock.

———. 1993. Compensation: what does it mean? *TaimLain* 1(1):57-62.

———. 1995. 'Ritual movements reconsidered: Ethnohistory in Aluni' in *Papuan borderlands*. Edited by A. Biersack. Ann Arbor: University of Michigan Press. pp. 87-110.

Strathern, Marilyn. 1988. *The gender of the gift: Problems with women and problems with society in Melanesia*. Berkeley: University of California Press.

———. 1997. Multiple perspectives on intellectual property. Unpublished paper presented at the Seminar on Intellectual, Biological and Cultural Property Rights, August, Port Moresby.

Strehlow, T. 1970. 'Geography and the totemic landscape in central Australia: A functional study' in *Australian Aboriginal anthropology.* Edited by R. Berndt. Nedlands: University of Western Australia Press.

Stürzenhofecker, G. 1994. Visions of a landscape: Duna pre-meditations on ecological change. *Canberra Anthropology* 17(2):27-47.

Sullivan, P. 1997. 'Dealing with native title conflicts by recognising Aboriginal political authority' in *Fighting over country: Anthropological perspectives*. Edited by D. Smith and J. Finlayson. Canberra: Centre for Aboriginal Economic Policy Research, Australian National University. pp. 129-40.

Sutton, Peter 1978. Aboriginal society, territory and language at Cape Keerweer, Cape York Peninsula, Australia. PhD thesis, University of Queensland.

———. 1988. 'Myth as history: History as myth' in *Being black: Aboriginal cultures in 'settled' Australia*. Edited by Ian Keen. Canberra: Aboriginal Studies Press. pp. 25-268.

———. 1995. 'Atomism versus collectivism: The problem of group definition in native title cases' in *Anthropology in the native title era*. Edited by J. Fingleton and J. Finlayson. Canberra: Aboriginal Studies Press. pp. 1-10.

———. 1996. The robustness of Aboriginal land tenure systems: Underlying and proximate customary titles. *Oceania* 67(1):7-29.

Sutton, P., and B. Rigsby. 1982. 'People with "politicks": Management of land and personnel on Australia's Cape York Peninsula' in *Resource managers: North American and Australian hunter-gatherers*. Edited by N. Williams and E. Hunn. Boulder: Westview Press. pp. 155-71.

Swain, Tony. 1993. *A place for strangers: Towards a history of Australian Aboriginal being*. Melbourne: Cambridge University Press.

Taussig, M. 1980. *The Devil and commodity fetishism in South America*. Chapel Hill: University of North Carolina Press.

———. 1987. *Shamanism, colonialism, and the wild man: A study in terror and healing*. Chicago: University of Chicago Press.

Taylor, J. 1984. Of acts and axes: An ethnography of socio-cultural change in an Aboriginal community Cape York Peninsula. PhD thesis, James Cook University.

Thomason, B.C. 1982. *Making sense of reification: Alfred Schutz and constructionist theory*. London: MacMillan Press.

Thomson, D.F. 1949. *Economic structure and the ceremonial exchange cycle in Arnhem Land*. Melbourne: Macmillan.

———. 1957. Early Macassar visitors to Arnhem Land and their influence on its people. *Walkabout*. 23(7):29-31.

Tindale, N. 1974. 'Australia north east: Tribal boundaries map' in *Aboriginal tribes of Australia*. Berkeley: University of California Press.

Toft, S. (ed.). 1997. *Compensation for resource development in Papua New Guinea*. Canberra: Law Reform Commission of Papua New Guinea et al.

Tonkinson, R. 1991. *The Mardu Aborigines: Living the dream in Australia's desert*. 2nd ed. Fort Worth: Holt Rinehart Winston.

———. 1997. Anthropology and Aboriginal tradition: The Hindmarsh Island bridge affair and the politics of interpretation. *Oceania* 68(1):1-26.

Trigger, D. 1992. *Whitefella comin': Aboriginal responses to colonialism in northern Australia*. Cambridge: Cambridge University Press.

———. 1997. 'Reflections on Century Mine: Preliminary thoughts on the politics of Aboriginal responses' in *Fighting over country: Anthropological perspectives*. Edited by D. Smith and J. Finlayson. Canberra: Centre for Aboriginal Economic Policy Research, Australian National University. pp. 110-28.

———. 1998. 'Citizenship and indigenous responses to mining in the Gulf Country' in *Citizenship and indigenous Australians: Changing conceptions and possibilities*. Edited by N. Peterson and W. Sanders. Cambridge: Cambridge University Press. pp. 154-66.

Turner, T. 1988. 'Ethno-ethnohistory: Myth and history in native South American representations of contact with western society' in *Rethinking history and myth. Indigenous South American perspectives on the past*. Edited by J.D. Hill. Chicago: University of Illinois Press. pp. 235-81.

———. 1998. Anthropological activism, indigenous peoples and globalisation. Paper prepared for the Society for Applied Anthropology, San Juan, Puerto Rico, April 1998.

Vail, J. 1995. 'All that glitters: The Mt Kare gold rush and its aftermath' in *Papuan Borderlands: Huli, Duna, and Ipili Perspectives on the Papua New Guinea Highlands*. Edited by A. Biersack. Ann Arbor: University of Michigan Press.

Verdery, Kathryn 1992. 'The "etatization" of time in Ceausescu's Romania' in *The politics of time*. Edited by Henry J. Rutz. American Ethnological Society Monograph Series, No. 4. Washington, DC: American Anthropological Association.

Von Sturmer, J. 1978. The Wik region: Economy, territoriality and totemism in western Cape York Peninsula, North Queensland. PhD thesis, University of Queensland.

Wagner, Roy. 1967. *The Curse of Souw*. Chicago: University of Chicago Press.

———. 1972. *Habu: The innovation of meaning in Daribi religion*. Chicago: University of Chicago Press.

———. 1974. 'Are there social groups in the New Guinea Highlands?' in *Frontiers of anthropology*. Edited by M.J. Leaf. New York: D. Van Nostrand Co.

———. 1975. *The invention of culture*. Englewood Cliffs, NJ: Prentice-Hall.

———. 1998. The production of human focality. *Social Analysis* 42(3):55-66.

Warner, W.L. 1932. Malay influence on the Aboriginal cultures of north Australia. *Oceania*. 2(4):476-95.

———. 1958/1969. *A black civilization. A social study of an Australian tribe*. Chicago: Harper & Roe.

Weiner, J. 1988. *The heart of the pearl shell: The mythological dimension of Foi sociality*. Berkeley: University of California Press.

———. 1991. *The empty place*. Bloomington: Indiana University Press.

———. 1994. The origin of petroleum at Lake Kutubu. *Cultural anthropology* 9:37-57.

———. 1995a. *The lost drum: The myth of sexuality in Papua New Guinea and beyond.* Madison: University of Wisconsin Press.

———. 1995b. Technology and *techne* in Trobriand and Yolngu art. *Social Analysis* 38:32-71.

———. 1998. Afterword: Revealing the grounds of life in Papua New Guinea. *Social Analysis* 42(3):135-42.

———. 1999. Culture in a sealed envelope: The concealment of Aboriginal heritage and tradition in the Hindmarsh Island Bridge affair. *Journal of the Royal Anthropological Institute* 5(2):193-210.

———. n.d. Religion, belief and action: The case of Ngarrindjeri 'women's business' on Hindmarsh Island, South Australia, 1994-1996. MS in author's possession.

Welsch, R. 1987. Multinational development and customary land tenure: The Ok Tedi project of Papua New Guinea. *Journal of Anthropology* 6(2):109-54.

———. 1991. 'Ningerum' in *Encyclopedia of world cultures.* Vol 2, *Oceania.* Edited by Terence E. Hays Boston: G.K. Hall. pp. 245-8.

Westermarck, G. 1997. Clan claims: Land, law and violence in the Papua New Guinea Eastern Highlands. *Oceania* 67:218-33.

Williams, F.E. 1977. 'Natives of Lake Kutubu, Papua' in *The Vailala madness and other essays.* Edited by E. Schwimmer. Honolulu: University of Hawai'i Press.

Williams, N. 1986. *The Yolngu and their land: A system of land tenure and the fight for its recognition.* Canberra: Australian Institute of Aboriginal Studies.

Wilson, Rob, and Wimal Dissanayake. 1996. 'Introduction: Tracking the global/local' in *Global/local-cultural production and the transnational imaginary.* Edited by Rob Wilson and Wimal Dissanayake. Durham, NC: Duke University. pp. 1-18.

Wolfers, T. 1992. 'Politics, development and resources: Reflections on constructs, conflict, and consultants' in *Resources, development and politics in the Pacific Islands.* Edited by S. Henningham and R.J. May. Bathurst: Crawford House Press.

Wood, A.W. 1984. Land for tomorrow: Subsistence agriculture, soil fertility, and ecosystem stability. PhD thesis, University of Papua New Guinea.

Wood, Michael. 1995. 'White skins', 'real people' and 'Chinese' in some spatial transformations of the Western Province, PNG. *Oceania* 66:23-50.

———. Forthcoming. Logging, women and submarines: Some changes in

Kamula men's access to transformative powers. *Oceania.*

Wordick, F.J.F. 1982. *The Yindjibarndi language.* Pacific Linguistics Series C, No. 71. Canberra: Australian National University.

Worsley, P.M. 1955. Early Asian contacts with Australia. *Past and Present.* 7:1-11.

Zimmer-Tamakoshi, L. 1993. 'Bachelors, spinsters, and *pamuk meris*' in *The business of marriage: Transformations in Oceanic matrimony.* Edited by R. Marksbury. Pittsburgh: University of Pittsburgh Press.

———. 1997. When land has a price: Ancestral gerrymandering and the resolution of land conflicts at Kurumbukare. *Anthropological Forum* 7(4): 649-66.

Zorc, R. David. 1986. *Yolngu-Matha dictionary.* Batchelor: School of Australian Linguistics, Darwin Institute of Technology.

INDEX

www.ingramcontent.com/pod-product-compliance
Lightning Source LLC
Chambersburg PA
CBHW030717250326
R18027900001B/R180279PG41599CBX00019B/29
* 9 7 8 0 9 5 4 5 5 7 2 3 2 *